George Vernon Hudson

An elementary manual of New Zealand entomology

Being an introduction to the study of our native insects

George Vernon Hudson

An elementary manual of New Zealand entomology
Being an introduction to the study of our native insects

ISBN/EAN: 9783337276089

Printed in Europe, USA, Canada, Australia, Japan

Cover: Foto ©berggeist007 / pixelio.de

More available books at **www.hansebooks.com**

AN ELEMENTARY

MANUAL

OF

NEW ZEALAND ENTOMOLOGY.

BEING

An Introduction to the Study

OF

OUR NATIVE INSECTS.

WITH 21 COLOURED PLATES.

BY

G. V. HUDSON, F.E.S.,

WELLINGTON, NEW ZEALAND.

London:

WEST, NEWMAN, & CO., 54, HATTON GARDEN.

1892.

To

The Right Hon. LORD WALSINGHAM,

M.A., F.R.S., F.L.S., F.Z.S.,

LATE PRESIDENT OF THE ENTOMOLOGICAL SOCIETY OF LONDON,

THIS LITTLE BOOK IS RESPECTFULLY DEDICATED

BY THE AUTHOR.

PREFACE.

THE object of the present volume is to give a brief account of the Natural History of the insects inhabiting New Zealand in a form intelligible to the ordinary reader. For this reason every effort has been made to avoid all unnecessary technicalities, and to adapt the book as far as possible to the requirements of youthful entomologists and collectors.

Several very elaborate systematic lists and descriptions have been published from time to time of the insects of New Zealand, amongst which may be specially mentioned—Captain Broun's "Manual of New Zealand Coleoptera," the illustrated "Catalogue of New Zealand Butterflies," edited by Mr. Enys, and Mr. Meyrick's "Monographs" of various groups of the Lepidoptera; but as yet no attempt has been made to present the subject in a suitable form for beginners.

It is hoped that this book will, to some extent, fill up the blank, and help to render what is now one of the most popular natural sciences in Europe, equally appreciated in New Zealand.

The author is much indebted to Captain Broun, Mr. R. W. Fereday, Mr. E. Meyrick, and others, for assistance in identifying the various species mentioned in this work.

Wellington, New Zealand, 1891.

CONTENTS.

ELEMENTARY MANUAL

OF

NEW ZEALAND ENTOMOLOGY.

CHAPTER I.

General Observations.

IN the present chapter I propose to give a brief sketch of the general principles of Entomology, including a rudimentary glance at the anatomy and classification of insects ; after which I think the reader will be in a better position to study the habits and life-histories of the individual species which follow.

The first requisite is a definition of what constitutes an INSECT.

An Insect is an articulate animal having the body divided into three distinct divisions, viz., the HEAD *(Fig. I. A), the* THORAX *(B), and the* ABDOMEN *(C). It is furnished with three pairs of legs, and generally has two pairs of wings, and to acquire this structure the creature passes through several changes, termed its metamorphoses.*

The head exhibits no distinct divisions, but bears the following appendages : the eyes, antennæ, and organs of the mouth, or trophi.

The eyes are of two kinds, compound and simple. The former (Fig. I. c c) are situated on the sides of the head above the mouth, and consist of two large hemispheres, composed of a great number of hexagonal divisions, each of which is a complete eye in itself. The latter (s s) are usually three in number, and are situated on the top of the head between the compound ones. They are, however, frequently wanting.

The antennæ (a) are two jointed organs, one of which is placed on each side of the head, between the eyes ; their functions are at present extremely doubtful, but they are invariably found in all insects.

The organs of the mouth consist of the following : the labrum (Fig. II. 3), or upper lip, a horny plate, closing the mouth from above ; the mandibles (1 1), or upper jaws, two strong bent hooks, articulated to the head on each side of the mouth, and opposed to one another like scissor blades ; the maxillæ (2 2), or under jaws, resembling the mandibles, but more delicately constructed, and furnished with a pair of jointed appendages termed maxillary palpi (5 5) ; and the labium (4), or lower lip, consisting of a horny plate somewhat resembling the labrum, but provided with two jointed appendages termed the labial palpi (6 6). All these organs are subject to great modification in suc-torial insects, which I shall notice further on, when dealing with the differences between the various orders.

The thorax consists of three primary divisions, *viz.,* the prothorax (Fig. I. b), mesothorax (d), and metathorax (k). The upper surfaces of these are termed the pronotum, mesonotum, and metanotum respectively, and the under the prosternum, mesosternum, and metasternum ; other divisions exist in some insects, but they are not of a sufficiently

general character to be noticed here. The six legs are attached to the under surface of the thorax, a single pair to

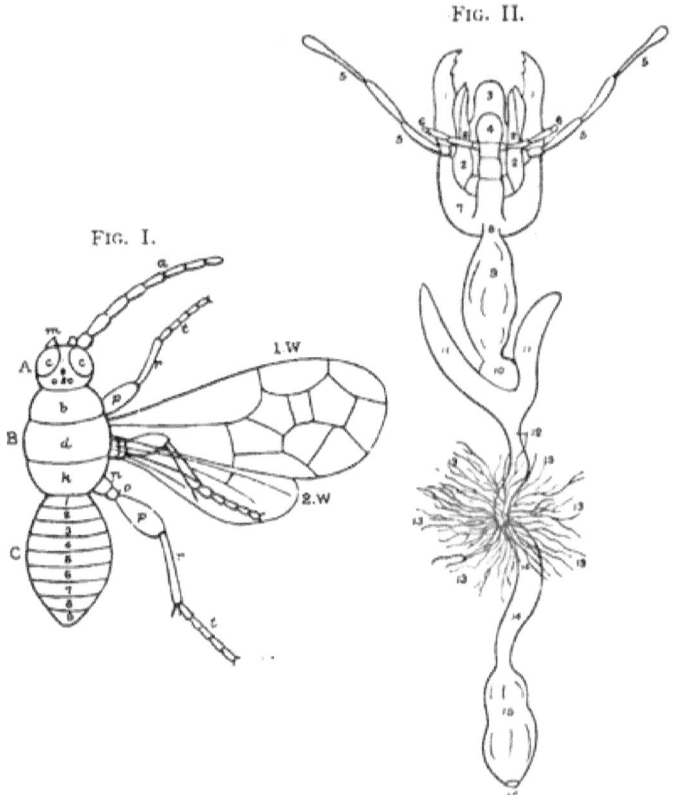

FIG. II.

FIG. I.

FIG. I.—Body of an insect (Hymenoptera), showing the principal divisions :
A, head ; B, thorax ; C, abdomen ; *a*, antenna ; *c*, compound eyes ; *m*,
mandible ; *s*, simple eyes ; *b*, prothorax ; *d*, mesothorax ; *k*, metathorax ;
1W, fore-wing ; 2W, hind-wing ; *n*, coxa ; *o*, trochanter ; *p*, femur ; *r*,
tibia ; *t*, tarsus ; 1 to 9 segments of the abdomen.

FIG. II.—Oral and digestive system of *Deinacrida megacephala* (this insect is
drawn on Plate XVIII., fig. 2) : 1, mandibles ; 2, maxillæ ; 3, labrum ; 4,
labium ; 5, maxillary palpi ; 6, labial palpi ; 8, œsophagus ; 9, crop ; 10,
gizzard ; 11, pancreas ; 12, stomach ; 13, biliary vessels ; 14, ilium ; 15,
colon ; 16, anus.

each division ; they are composed of the following joints :
coxa (Fig. I. n), trochanter (o), femur (p), tibia (r), and
tarsus (s).

The wings are attached to the meso- and metanotum; they consist of two membranes traversed by numerous horny ribs (Fig. I. 1W and 2W).

The abdomen is made up of nine segments (C 1 to 9), some of which are not infrequently wanting. It contains the organs of nutrition, circulation, and generation.

The digestive system, the structure of which is apparent from Fig. II., consists of the following divisions : the throat, or œsophagus (8) ; the crop (9) ; the gizzard, or proventriculus (10) ; the pancreas (11 11) ; the stomach, or ventriculus (12) ; the biliary vessels (13 13 13) ; the ilium, or little gut (14 14) ; and the colon (15) ; ending in the anus (16). In the suctorial tribes, the crop is modified into a very peculiar organ, termed the sucking stomach, which presents itself as a small bag, attached to the throat by a thin tube. This bag exhausts the air from the throat, when the insect is sucking, thus producing a vacuum therein, and causing a rapid ascent of fluid into the stomach.

The heart of insects consists of an elongated tube lying along the back, and termed the dorsal vessel. It is composed of a variable number of chambers, the blood being driven forward towards the head by its contractions. These motions may be easily seen in transparent species.

The breathing organs are distributed throughout the body in the form of numerous minute air-tubes, which are supplied with air from a variable number of apertures, situated on the sides of the insect, and termed spiracles.

The nervous system consists of a chain of ganglia, running down the ventral surface of the insect, and analogous to the spinal cord of higher animals. The number of ganglia varies greatly among the different tribes.

The metamorphosis of insects, which I have previously mentioned as one of their most essential attributes, consists of four distinct stages, *viz.*, the Egg, Larva, Pupa, and Imago.

The eggs of these animals exhibit a great diversity in shape among the different species. They are deposited by the parent with unerring instinct on substances suitable for the food of the larvæ, which, in the majority of cases, is quite different from that on which she herself subsists.

The larva state immediately succeeds the egg, and is spent almost exclusively in feeding, the insect growing at a great rate, and being frequently compelled to change its skin.

The pupa is usually completely quiescent, the insect being at this time quite incapable of any motion, except, perhaps, a slight twirling of its abdomen. Exceptions to this rule occur, however, in two of the orders, in which the pupa state does not differ materially from that preceding it.

In the imago, or perfect state, the insect appears under its final form, with every organ completely developed.

We will now consider the seven great divisions, or Orders, into which insects are divided, the complete knowledge of which is one of the most important elements in the ento- mologist's preliminary education. I trust that by a care- ful perusal of the following definitions, aided by references to the Plates, which illustrate numerous members of each order in their several states, the reader will be enabled to master the subject without much difficulty.

Order I.—COLEOPTERA.

Wings four ; the anterior pair (termed elytra) horny and opaque, the posterior membranous, and employed in flight ; mouth masticatory. The larva a grub with or without legs, but a distinct head always present. The pupa inactive, taking no food, the limbs of the future insect enclosed in distinct cases, and applied closely to the body. This is the largest of the Orders, and consists of all those insects popularly known as Beetles. (Plates I. and II.)

Order II.—HYMENOPTERA.

Wings four, membranous, the posterior pair being the smaller, and connected with the anterior during flight by a row of minute hooklets ; mouth masticatory, the maxillæ and labium being elongated, in many of the families, into a long sucking instrument or "tongue." Metamorphosis as in the Coleoptera. A large Order, containing the numerous tribes of Sawflies, Bees, Wasps, Ants, and Ichneumon-flies. (Plate III.)

Order III.—DIPTERA.

Wings two ; the posterior pair represented by two minute clubbed appendages termed poisers ; mouth a suctorial tube formed by an elongation of the labium, enclosing within it a variable number of setæ answering to the mandibles, &c., of biting insects. The larva without legs, a distinct head being often absent. The pupa inactive, the limbs of the imago firmly attached to the body, but plainly visible. Among the majority of species included in this Order the larval skin is not cast away, but envelopes the insect in a hard shell ; the true pupa is consequently only visible on the removal of this covering, when it is found to closely resemble those in which no such arrangement occurs. The Order comprises the numerous Gnats and two-winged Flies. (Plates IV., V., VI., VII.)

Order IV.—LEPIDOPTERA.

Wings four, generally covered with scales ; the anterior pair slightly superior in size ; mouth suctorial, the maxillæ forming a spiral tongue, which is coiled between the large labial palpi when not in use; other oral organs rudimentary. In many instances the whole mouth and alimentary canal are more or less obliterated, a considerable number of the species taking no food in their final state. The larvæ always possess a distinct head and six thoracic legs, and in addition a variable number of prolegs are often present on the abdominal segments. Pupa inactive, the limbs of the future insect being usually indicated by lines in the integment. This Order contains all the varied tribes of Butterflies and Moths. (Plates VIII., IX., X., XI., XII., XIII.)

ORDER V.—NEUROPTERA.

Wings four, of equal size, membranous, and traversed with numerous branching ribs; the mouth masticatory, and in many instances but slightly developed. Larva with a distinct head and three strong thoracic legs; chiefly carnivorous. Pupa inactive; the limbs very perceptible and loosely applied to the body, but incapable of distinct motion. A small Order, comprising the Stoneflies, Lace-wings, Ant-lions, &c. (Plate XIV.)

ORDER VI.—ORTHOPTERA.

Wings four, of nearly equal size; the anterior pair often more or less leathery, but with distinct veins. The larva and pupa closely resembling the imago; the latter with rudimentary wings. In the instances where these organs are wanting in the mature insect, the metamorphosis merely consists of a series of moultings, and it is consequently a matter of some difficulty to determine when the insect is full-grown. This Order is of small extent; it includes the Earwigs, Cockroaches, Grasshoppers, Crickets, Termites, Dragonflies, Mayflies and Perlidæ; the last four being transferred from the Neuroptera of most authors. The minute species of Mallophaga and Thy-sanura will also come under this heading. (Plates XV., XVI., XVII., XVIII., XIX.)

Order VII.—HEMIPTERA.

Wings four, in some cases wholly membranous, but in a large proportion of the families the basal portions of the anterior pair are horny, and form protective cases for the other pair when not in use ; mouth suctorial, consisting of an elongate rostrum, enclosing four fine setæ. The larva and pupa resemble the imago, the latter being active, with rudimentary wings. In a few instances, a slight divergence from the parent form is shown in the preparatory states (Cicadas, &c.). This is a small Order, containing the Cicadas or " Singers," Bugs, Plant Lice, and all the suctorial animal lice. (Plate XX.)

After the Orders, the divisions to be considered are the Groups, Families, Genera and Species.

Groups are large divisions immediately subordinate to the orders, and consist of a number of *kindred* families. They are of great assistance to the student in dealing with the very large Orders, such, for instance, as the Coleoptera.

Families, again, consist of a number of allied genera, and Genera, in the same way, of allied species.

With regard to the Families, I have in the main followed those of Professor Westwood in his ' Modern Classification of Insects,' as most recent writers appear very much divided in opinion as to the correct limits of these divisions. Much diversity also prevails with respect to the proper definitions of Genera and even Species, but I have deemed it best to follow the authority of the latest catalogues in this matter, as any changes in nomenclature are always liable to produce confusion.

CHAPTER II.

Collecting Insects.

So many excellent essays have been written on collecting
insects that it would probably be a most difficult task to
supply much fresh information on the subject ; but as many
of my readers may be unable to consult works specially
devoted thereto, the present chapter will, perhaps, be of
some value in showing them a few of the most convenient
methods of collecting insects in New Zealand.

Coleoptera, or Beetles, may be found almost everywhere.
Overturning logs and stones, peeling off bark, and cutting
into the solid wood of trees, all produce a great variety of
species. A small axe and an iron wrench, shaped some-
thing like a chisel, but bent round at the upper end, are
the best instruments for working old trees. The bark should
be all stripped off and examined, as well as the surface of
the log underneath. The same remarks apply to stones,
which should be searched as well as the places from which
they were removed. Sacks, if left about the fields for a
few weeks, often harbour good beetles, and when found
they should always be pulled up and examined.

An umbrella, held upside down under flowering shrubs
in the forest, will often be found swarming with beetles
after the plants have been sharply tapped with a stout

walking-stick. The same object may be attained by
spreading a newspaper, or sheet, under the trees and
then shaking them ; the beetles will fall on to the sheet,
and may then be captured. The only advantage of the
umbrella is that it can be more readily used in awkward
places, such as on steep hill sides.

The dead bodies of birds and animals also contain
peculiar species ; they may be held over the umbrella and
shaken into it, when the inhabitants will fall out, and can
easily be obtained. Dead fish on the sea beach are often
very productive. Moss and fungi are unfailing resorts
of many of the smaller species of Coleoptera, and can be
examined in the winter when the entomologist is otherwise idle.

Beetles should always be brought home alive. The
small round tin boxes sold with Bryant and May's wax
matches will be found very serviceable for this purpose.
These boxes are far better for all kinds of collecting than
either pill- or chip-boxes, as they do not break when knocked
about. A separate box should always be given to a large
or rare species, but most of the smaller kinds will travel
quite safely in company, especially if a wisp of grass or a
leaf is put into the box to give them foothold.

Beetles must be killed with boiling water, and left
immersed some hours before setting. They must be
pinned through either the right or left elytron, and each
collector must always keep to one side, as nothing looks
worse than to see some of the specimens pinned on the
right and others on the left side. When pinned the beetles
are set on a corked board, the legs, &c., being placed in a
natural position, and retained until dry by means of pins
and pieces of paper and card. The smaller species should
be mounted with transparent gum on a neat piece of card,
which can be pinned in the store-box or cabinet with
the others. The greatest care should be taken to set

symmetrically, so that the limbs on the right-hand side of an insect are in the same position as those on the left.

Hymenoptera may be captured with the ordinary butterfly-net, and are found abundantly during the summer. The larger species are pinned through the centre of the thorax, and set in the same way as Coleoptera, the smaller ones on card with gum. These insects should, if possible, be made to fly into the vessel of boiling water, as by this means they generally die with their wings expanded, which is a great assistance when setting them. This can usually be managed by holding the box containing the specimen immediately over the water, and giving it a sharp tap with the finger of the other hand.

Diptera are also captured with the net, and pinned in the same way, but should be killed with the laurel bottle.

Lepidoptera are the most difficult of all to collect, and are at the same time the most attractive to beginners. They may be captured with a net made of fine gauze (mosquito net dyed green is the best material) ; the frame to support the net is constructed of a piece of cane bent into a hoop, each of the ends being supported in a forked tube shaped like a Y, and the long tube, forming the base of the Y, is firmly fitted on to the end of a walking-stick. This form of net is light, strong, and easily made ; the only thing requiring special attention is the Y, but this can be readily made by any tinsmith out of two pieces of gas-pipe of different sizes, the larger one for the stick, and the smaller one for the ends of the cane to fit into. The collector should also be furnished with a number of small tin boxes.[1] All this apparatus can easily be packed into an ordinary satchel.

[1] For Lepidoptera I can strongly recommend "Jahncke's Patent Round Boxes" with glass lids. They may be obtained from any chemist, or from Messrs. Sharland & Co., Wholesale Druggists, Wellington.

When the entomologist reaches his hunting-ground, he
will mount his net and place a number of the boxes in
his left-hand coat pocket. The foliage of all trees and
shrubs should be vigorously beaten and the insects captured as they fly out. When a moth is taken, the collector
will first turn the net half way round so as to close the
entrance, and then, directly the insect ceases fluttering, he
should carefully place one of the little boxes over it and
slip on the lid. The box is then transferred to the right-
hand pocket. He will soon learn to do this without in any
way damaging the insect. On arrival at home, the insects
should be immediately killed in the laurel bottle. This is
an ordinary wide-necked bottle with a small bag of well-
bruised *young* laurel shoots at the bottom, covered with a
circular piece of card fitting accurately to the sides of the
bottle. Laurel shoots can always be obtained about the
middle of October, when several killing bottles can be prepared. They must always be wiped out before using, and
kept carefully corked. After a few hours the insects
should be tilted out of the bottle on to a tablecloth, and
pinned exactly through the centre of the thorax. The
rough surface of the tablecloth prevents them from slip-
ping during the operation. About one-third of an inch of
pin should project below the body of the insect. If a moth
or butterfly dies with its wings folded upwards over the
back, it must be carefully picked up between the thumb
and index finger of the left hand, and the pin inserted with
the corresponding fingers of the right hand. When all are
pinned they should be transferred to a tin box, lined with
cork, which has been previously well damped with water.
While pinning them into this box great care must be taken
not to allow the wings to come in contact with the damp
cork. In about twenty-four hours the specimens thus treated
will be ready for setting. This process is performed by
means of corked boards of various widths for different sized

species. Each board has a groove down the centre for the
bodies of the insects to rest in, while the wings are spread
out on either side. They should be carefully moved forwards with a fine-pointed needle to the desired position,
and retained by strips of tracing cloth pinned firmly down
at the ends. These strips must not be removed until the
insects are thoroughly dry and ready to place in the store-
box or cabinet. In setting Lepidoptera, as with other
insects, symmetry and a natural position are the main
points to be aimed at, special care being taken that the
antennæ, fore- and hind-legs, and wings, are shown in
correct positions, the middle pair of legs being of course,
in the majority of cases, hidden by the wings. It is almost
needless to say that different sized pins should be used for
various insects, but this point must be left to the discretion
of the collector. Entomological pins of all sizes can be
obtained from James Gardner, of 29 Oxford Street, London.
Gilt pins are useful for many species which are liable to
form verdigris on the pins, and are universally employed
by many entomologists, but are probably not so strong as
the silvered ones.

Many species of moths are only to be found at night.
When working at this time the collector must suspend a
bulls-eye lantern round his neck or waist, and can then have
both arms free for capturing insects on the wing or at
blossoms. Honey mixed with a little rum, and applied
with a small brush to the trunks of trees a few minutes
after sunset, will, on some evenings, attract large numbers
of valuable species, but not infrequently it is quite unpro-
ductive. This mode of collecting has been termed "sugar-
ing" by entomologists, and may be employed during the
whole summer. The best blossoms for attracting insects
in New Zealand are those of the white rata,[1] which blooms
in the forest from February till April, and from which the

[1] Metrosideros scandens.

collector may generally rely on getting a rich harvest. The insects can usually be slipped directly from the flowers into the killing bottle.

This is much better than netting them, although occasionally one will escape during the process. When dead the specimens should be placed in a small tin box which has been filled with cotton-wool, packed very lightly. In this way a large number of moths may be carried a long distance with perfect safety, and the extremely inconvenient process of pinning them in the field obviated. If Jahncke's patent boxes are employed it is quite unnecessary to kill the moths in the field. They can be boxed directly from the blossoms and taken home alive without suffering any injury.

Lepidoptera, and in fact all insects, are attracted by light, and in some situations the collector will find that he may frequently obtain good species by merely opening his sitting-room window and waiting for the insects to arrive. Much of course depends on the situation of the collector's residence and the nature of the night, which should be dark and warm. I have occasionally tried taking a lamp into the forest to attract insects, but have not met with much success. In swampy and flat situations, no doubt, attracting by light would be very effective, especially if a powerful lamp was employed, in an exposed situation, with a sheet behind it, supported between two poles. This method has been followed with great success by many English entomologists in the fens, but has not yet been tried in the New Zealand swamps, where it would probably be the means of bringing many new and interesting species to " light."

With regard to collecting members of the three remaining Orders but little need be said. Neuroptera can be treated in the same way as Lepidoptera, but they should be set on flat boards. The treatment of the Orthoptera will resemble that of the Coleoptera, but the larger species will require

to be stuffed with cotton-wool before setting. A few of
the largest species of the Lepidoptera must also be stuffed.
For this purpose the specimens should be placed on their
backs on a piece of clean glass so that none of the scales
may be rubbed off. After the contents have been removed,
a little chalk should be introduced into the abdomen with
the cotton-wool. Hemiptera can be collected and set like
Coleoptera, but some of the more delicate species, such as
the *Cicadæ*, should be killed in the laurel bottle instead of
in boiling water.

Before concluding the present chapter I should like to
say a few words on the subject of rearing insects, which the
entomologist will soon learn to regard as by far the most
interesting method of acquiring specimens for his collection.

Members of the Coleoptera are probably the most difficult
insects to rear in captivity. Their larvæ may be kept in
ordinary jam-pots covered with perforated zinc, and filled
with earth or rotten wood. The carnivorous species must,
of course, be supplied with the animals on which they feed.
Beetle larvæ are often some years in attaining maturity.
Many of the Hymenoptera and some of the Diptera are
parasitic on the larvæ of the Lepidoptera ; they are consequently found in rearing these insects, and their economy
should always be carefully recorded.

Lepidoptera are, perhaps, the most satisfactory insects
to rear. Most of the larvæ feed on the leaves of different
plants, and all that is needed is to keep them well supplied
with fresh food.

So great a variety of cages have been devised for the
rearing of caterpillars that it would be quite impossible to
describe them here. I will therefore only give a short account
of those which I have used myself, and have found so convenient that I do not hesitate in recommending them to
those entomologists who wish not only to rear insects but
to study their habits.

The cages I have been in the habit of using are made of two or three thicknesses of cardboard bent round into a cylinder and strongly pasted together. They may be of various sizes, from three to four inches in diameter up to eight or ten, and constructed so that one will go inside the other. The height should exceed the diameter by about one and a half inches. The cylinders should be made so as to stand exactly level on a flat surface, and they should have two rows of small openings round the sides for the admission of air. It is a good plan to have four of these openings in each row and place them opposite one another. They should be covered on the inside with gauze, stiffened with green or brown paint, as the dark colour will enable the observer to see inside more readily. A circular piece of glass is fitted into the upper end of the cylinder, and fixed by means of paste and paper. The base of the cage consists of two round pieces of wood, one about half an inch smaller than the other, the smaller one nailed exactly in the centre of the larger piece. These are made so that the cardboard cylinder fits *accurately* on the outside of the smaller piece of wood. The whole cage is then neatly covered with white paper inside and brown outside. A complete view of the interior can of course be obtained by looking in at the top, while the cages can be stowed away one within the other when not in use. A stone ink-bottle should be put on the floor of each cage and filled with water, into which a sprig of the food-plant can be introduced. Care must be taken to plug up the mouth of the bottle, so that the larvæ may not crawl down the stem of the plant into the water and thus meet with an untimely end. This may readily be done by means of a cork with a hole bored in it for the stem to pass through, or a plug of moss or blotting-paper. Members of almost all the orders can be reared in these cages, as jam-pots full of earth may easily be introduced, in the place of the stone

bottle, when required for species which bury. A circular piece of blotting-paper should be placed over the
bottom of each cage, while larvæ are feeding in them, and
renewed when at all soiled. The excrement must also be
removed when the larvæ are supplied with fresh food.
As a rule, this is only necessary about twice a week, as the
water will keep most plants fresh for quite a lengthened
period. When it is necessary to remove a larva it should
always be done with a fine camel-hair brush, never with
the fingers. Generally, however, it is better to allow the
larvæ themselves to crawl from the old sprig on to the new
one, which they usually do in a few hours after the food is
changed. The old plants should of course then be taken
out so as to afford more room for fresh air.

Many female moths may be induced to lay their eggs
in captivity, especially if put in a box with some of the
food-plant of the larva. It is extremely instructive and
interesting to rear an insect from the egg. When the
young larvæ first emerge they must be kept in a tumbler
with a piece of glass put over the top, as they might escape
through the ventilators of the cages, but they ought to be
transferred immediately they are large enough. When
rearing a lot of caterpillars from a batch of eggs, care
should be taken to avoid overcrowding.

A collection of insects should always eventually be
placed in a neatly constructed cabinet. They should be
arranged in rows, systematically, with the correct names
under each species, and the name of the order or group at
the commencement of each drawer. Numerous modifica-
tions in arrangement are often needed to meet the requirements of different sized insects, but an inspection of any
good collection will at once explain the general principles.
Camphor should be pinned in the corner of each drawer
or store-box, and the whole collection fumigated with
carbolic acid, or equal parts of oil of thyme, oil of anise,

and spirits of wine, every six months. These can be introduced in a watch glass containing a small quantity of the chemicals on a pellet of cotton-wool, care being taken not to stain the paper at the bottom of the drawer. For the same reason, while using carbolic acid, the camphor should be taken out, as otherwise it will "sweat." All boxes for the reception of insects must of course be lined with cork and paper.

It is most important that an accurate record should be kept of every specimen that is placed in the collection. This may be done by attaching to the pin underneath each insect a small numbered label, which refers to a book containing locality, date of capture and other particulars.

I have found it a good plan to give every species a number, and every specimen a letter. Thus, supposing *Vanessa gonerilla* is numbered "6," the first specimen taken would be "6a," the second "6b," and so on, all the specimens, perhaps, having different dates and localities. This system is very convenient when specimens are sent away to be identified by another entomologist, as, provided the collector always retains a single specimen of the species which he desires named, it obviates the necessity of having his specimens returned, the number showing at once to what species the name refers. At least five lines should be allotted to each species in the collection journal, and the writing should be small but distinct.

A collection formed in this manner will not only be a constant source of pleasure to the collector and those who succeed him, but very probably of great value in deciding many important questions in entomological science.

CHAPTER III.

The Coleoptera.

THE observations on the natural history of the New Zealand beetles, forming the subject of the present chapter, are much less numerous than might have been expected from the great number of species which have been described. The difficulties attendant on rearing these insects are, however, very great, and it thus happens that the life-histories here given bear a smaller proportion to the number of the Coleoptera than will be found to be the case with the majority of the other Orders. I hope, however, that the few details I have collected, referring to the following species, may induce some of my readers to investigate others for themselves.

Group GEODEPHAGA.

Family CICINDELIDÆ.

Cicindela tuberculata (Plate I., fig. 1, 1a larva).

This is a very abundant insect found throughout the country in all dry situations. It delights in hot sunshine, and may be constantly observed flying from our footsteps with great rapidity as we walk along the roads on a hot summer's day.

Its larva (Fig. 1a) is an elongate fleshy grub, the head

and first segment being horny and much flattened, and the body provided with two large dorsal humps, each bearing at its apex a slender curved hook.

The burrows of these insects are very conspicuous, and must have been noticed by every one, in garden paths, sandbanks, and other *dry* situations ; they are sometimes very numerous, and may be best described as perfectly round shafts, about one line in diameter, and extending to the depth of three or four inches, generally slightly curved at the bottom. The sides are perfectly smooth, and the larva may be often discovered near the mouth of its burrow, using its dorsal hooks to support it, and thus having both legs and jaws free to dispose of the unfortunate insects that fall into its snare. These usually consist of flies and small beetles, which appear to be urged by curiosity to crawl down these pitfalls, and thus bring about their own destruction. By reference to the figure it will be seen how admirably the hollowed head and prothorax serve the purpose of a shovel to the larva, when forming its shaft. These burrows are first observed about the middle of November ; the perfect insects coming abroad three weeks or a month later, when they may be often seen in the neighbourhood of their old domiciles. They are very voracious, devouring large quantities of flies, caterpillars, and other insects, some of which are much superior to themselves in size. On one occasion I saw a male speci-- men of *Cicindela parryi* (a species closely allied to but smaller than *C. tuberculata*) attack a large Tortrix cater-pillar, an inch and a half in length. The beetle invariably sprang upon the back of the caterpillar and bit it in the neck, being meanwhile flung over and over by the larva's vigorous efforts to free itself from so unpleasant an as-sailant. During the fight, which lasted fully twenty minutes, the beetle was compelled to retire periodically to gain fresh strength to renew its attacks, which were eventually suc--

cessful, the unfortunate tortrix becoming finally completely exhausted. The beetle devoured but a very small portion of the caterpillar, and abandoning the remainder went off in search of fresh prey. Eight other closely allied species of *Cicindela* are described by Captain Broun in the "Manual of the New Zealand Coleoptera," but they offer no especial peculiarities, and *C. tuberculata* may be taken as a type of the genus.

Family CARABIDÆ.

Pterostichus opulentus (Plate I., fig. 3, 3a larva).

This fine beetle is very common in most wooded situations in the Nelson district ; it may be at once distinguished from the numerous other closely allied species by the beautiful metallic coppery tints that adorn its thorax and elytra.

During the day it is usually discovered concealed under logs and stones, and when disturbed, rushes into the first crevice to get out of the light. At night time, it comes abroad to feed, killing an immense number of flies, caterpillars, and other insects, to satisfy its voracious appetite. Although of a most ferocious disposition, it is not wanting in maternal affection. The female, when about to deposit her eggs, excavates a small cavity nearly three inches square, in which they are placed. These she broods over until hatched, and probably some little time afterwards, as I have found a specimen close to a nest, which contained both eggs and larvæ, and the zealous mother furiously bit at anything presented to her. The eggs are oval in shape, quite smooth, and yellowish white in colour. The young larva is drawn at Plate I., fig. 3a ; it is remarkable for its superficial resemblance to a small Iulus, and being found in similar situations to that animal, its mimicry has probably some useful object. The older larva differs chiefly in having the head and thoracic segments proportionately smaller. Twenty-one closely allied insects belonging to two genera are described by Captain Broun in his Manual, the largest being *Pterostichus australasiæ*, which is found in similar localities to the present species, but is not so common.

Group HYDRADEPHAGA.

Family DYTICIDÆ.

Colymbetes rufimanus (Plate I., fig. 4, 4a larva).

This insect is found plentifully in all still waters during the summer months. Its larva is a soft elongate grub, provided with six slender thoracic legs, and a pair of powerful mandibles. The posterior extremity of the body is furnished with two curious appendages bearing a spiracle at the apex of each, which the larva frequently protrudes above the surface of the water. The air is taken in through the spiracles, and conveyed to all parts of the body by two main air-tubes, one of which springs from each spiracle, and branches throughout the insect in every direction. During the spring months the larvæ may be found of various sizes in similar situations to the imago ; they are very voracious, devouring freshwater shrimps, *Ephemera* larvæ, and occasionally, when pressed by hunger, they will even destroy individuals of their own species for food. These they capture by means of their powerful mandibles, retaining a firm hold of the victim until they have consumed all the fleshy portions, the rest of the carcase being thrown aside, and a fresh search made for more. One individual I kept for some time, remained perpetually concealed in a small patch of green weed, growing in the middle of its aquarium. In a short time it became surrounded with the skeletons of small water shrimps which had been seized by the larva as they passed by its hiding place, the unfortunate crustaceans only discovering their enemy when it was too late. I have not yet observed the pupa of this insect, but it probably does not differ materially from those of its European allies. Although so very different in general appearance to the preceding insects, this beetle will be found on careful examination to agree with them in all important respects, being only what a ground beetle might naturally become if forced to lead an aquatic existence. Breathing is effected in all the water beetles by the spiracles of the abdomen, which alone are developed. The air is taken in between the elytra and the body, and owing to the convexity of the former, a supply can be retained sufficient to last the insect some twenty or thirty minutes. The beetles may be often observed with the extremity of their elytra protruded above the surface, renewing their supplies of air. On very hot days *C. rufimanus* may be occasionally seen flying with great rapidity far away from its native ponds. When doing so it makes a loud humming noise, and is a much more conspicuous object than when in the water.

Group CLAVICORNIA.

Family NITIDULIDÆ.

Epuræa zealandica.

This curious little beetle is found abundantly in the neighbourhood of decaying fungi, throughout the year, being most plentiful in the autumn and early winter. Its larva is a small cylindrical grub, with the head and legs so minute that they are scarcely perceptible, causing it to closely resemble the maggots of many dipterous insects, occurring in similar localities. It is generally found in the large yellow fungi, so abundant in wet situations during the late autumn and winter months. It forms numerous galleries through the plant in all directions, and owing to the large amount of moisture which is usually present, these galleries are often filled with water, so that the insect may be said to be sub-aquatic in its habits. I have not yet detected the pupa of this species, although the discovery of a large quantity of both larvæ and perfect insects is of everyday occurrence with the entomologist in winter.

Family ENGIDÆ.

Dryocora Howittii (Plate I., fig. 6, 6a larva).

This quaint-looking little insect occurs occasionally in damp matai logs, when in an advanced state of decay. The larva (Fig. 6a) is very flat and thin, possessing the usual thoracic legs, which, however, are rather short. The last segment of the abdomen is furnished with an anal proleg and a pair of small setiform appendages. Its mode of progression is very peculiar, resembling that of the Geometer larvæ among the Lepidoptera.

The thoracic legs are first brought to the ground, and the rest of the body is then drawn up in an arched position close behind them. The anal proleg then supports the insect while the anterior segments are thrust out, and the others follow as before. This method is only employed on smooth surfaces, the larva crawling along elsewhere in the usual manner.

The perfect beetle is a very sluggish insect, and difficult to find owing to its colour, which closely resembles that of the wood in which it lives.

Family ENGIDÆ.

Chætosoma scaritides (Plate I., fig. 2).

This insect may be at once recognized by its peculiar shape, no other New Zealand beetle resembling it in this respect. Although tolerably common and generally distributed, it is very seldom seen abroad, spending almost the whole of its life concealed in the burrows of various wood-boring weevils. Its larva, which feeds on the grubs

of these insects, is of a pinkish colour, very fat and
sluggish ; the head and three anterior segments are strong
and horny, the legs being rather short. It undergoes its
transformation into the pupa within the weevil burrows,
when the limbs of the perfect insect can be seen folded
down the breast, the wings and elytra being much smaller
than in the beetle. Specimens in all stages of existence
may be readily procured by splitting up old perforated
logs which have been long tenanted by weevils.

Group BRACHELYTRA.

Family STAPHYLINIDÆ.

Staphylinus oculatus (Plate I., fig. 5).

This is the New Zealand representative of *S. olens* or
the " Devil's Coach Horse," one of the most familiar of
British beetles. It is found occasionally in the neighbourhood of slaughter-houses, and may be at once distinguished
from any of the allied species by a large spot of brilliant
scarlet situated on each side of its head behind the eyes ;
this very conspicuous feature has given it the specific name
of *oculatus*. I am at present unacquainted with the transformations of this fine insect, but they will probably closely
resemble those of the typical species (*S. olens*) described
in the majority of standard books on European Coleoptera. This beetle may be frequently seen flying in the
sunshine, when it has a most striking appearance, owing to
its large size and rapid motion. An unpleasant odour is
found to arise when it is handled, this being noticeable in
nearly all the members of the family. These beetles
are comparatively numerous in New Zealand, the genus
Philonthus comprising several elongate active insects,
of which *P. æneus* is one of the commonest, and may
be found abundantly amongst garden refuse. Others
frequent the seashore, feeding on decaying seaweed, and

may be noticed flying in all directions along the coast immediately after sundown. Another genus (*Xantholinus*) includes a number of interesting beetles found in old weevil burrows, and probably feeding on their inmates.

Group LAMELLICORNES.

Family LUCANIDÆ.

Dorcus punctulatus (Plate I., fig. 7).

An abundant species chiefly attached to the red pine tree or rimu, where it may be found concealed beneath the scaly bark, in the angles of the trunk near the roots. When disturbed, it folds up its legs and antennæ on its breast, and, extending its powerful jaws, awaits the approach of the enemy, ready to bite anything coming within its reach. These, however, are purely defensive measures, the insect being quite harmless when left alone. The larva is at present unknown to me. Another species, *D. reti-culatus,* is a much handsomer insect than the preceding; it may be at once recognized by four deep impressions in the thorax, filled in with light-brown scales; the margins of the elytra are similarly scaled, as well as four spots on each elytron, the remainder of the beetle being dark-brown and shining. It is generally found in totara bark, but is much scarcer than the last species. One small specimen I possess, remarkable for its brilliant appearance, was taken under the bark of a stunted black birch tree, over two thousand feet above the sea-level.

Family MELOLONTHIDÆ.

Stethaspis suturalis (Plate I., fig. 8, 8a larva).

This conspicuous insect occurs abundantly in all open situations. Its larva (Fig. 8a) inhabits the earth, feeding on the roots of various plants, and is especially abundant

in paddocks, where it occasionally does considerable damage
to the grass, and threatens ere long to become as great a pest
as its first cousin, the renowned Cockchaffer of England
(*Melolontha vulgaris*), whose fearful ravages need no description. It may be taken as a typical larva of the family,
the rest differing from one another in little else than size.
When full-grown it is quite as large as the illustration, and
is nearly always in the position there indicated, owing to
the size of its posterior segments and the absence of any
anal proleg, which compel it to lie always on its side. I
have not yet succeeded in obtaining the pupa of this insect,
although larvæ may be frequently found enclosed in oval
cells, evidently about to undergo their transformation.
Several of these have been kept in captivity, but they have
hitherto always died without undergoing any change. I
have, however, no doubt as to its being the larva of *S.
suturalis*, as there are no other large Lamellicorns found
near Wellington to which it could possibly be referred.
The perfect beetle appears in great numbers from November
to March ; it is best taken at dusk, when it flies with a
loud humming noise, about four feet above the ground. If
knocked down it always falls amongst the herbage, and is
not readily perceived until a few minutes later, when the
humming noise is resumed as the insect again gets under
weigh, and the would-be captor must not lose time if he
wishes to secure it. Occasionally individuals are seen
disporting themselves on the wing during the day, but
this must be regarded as a purely exceptional circumstance. Unlike the majority of nocturnal Coleoptera,
this insect does not appear to be attracted by light ; in
fact I have never obtained any specimens by this method,
although most other night-flying beetles may be taken in
goodly numbers at the attracting lamp.

Family MELOLONTHIDÆ.

Pyronota festiva.

This brilliant little insect is extremely abundant amongst manuka, during the early summer.   In general appearance it reminds one of a miniature specimen of the last species, but is more elongate in form ; the green thorax and elytra are also much brighter.   The latter are bordered with flashing crimson, the legs and under surface being reddish-brown, sparsely clothed with white hairs.   A small La-mellicorn grub, found amongst refuse in manuka thickets, is probably the larva of this insect ; it is less thickened posteriorly than that of *S. suturalis*, but otherwise closely resembles it.   The perfect insect is diurnal in its habits, flying round flowering manuka in countless numbers on a hot day.   The descent of thirty or forty of these little beetles on to the beating sheet, out of a single bush, is of frequent occurrence, and is particularly noticed by the New Zealand entomologist accustomed to the meagre supply of specimens offered in the majority of instances.

Group STERNOXI.

Family ELATERIDÆ.

Thoramus wakefieldi (Plate II., fig. 1, 1b larva, 1a pupa).

This fine beetle may be taken under rimu bark in tolerable abundance, and is often observed flying about at dusk during the summer.   Its larva inhabits rotten wood, usually selecting the red pine, in which it excavates numerous flat galleries near the surface of the logs.   When disturbed it is very sluggish, the head being immediately withdrawn into the large thoracic segment and completely concealed. The legs are very minute, and are of but little use in walking, the insect being chiefly dependent for locomotion on its large anal proleg, which is furnished with numerous horny spines.   When full-grown this larva closes up one end of

its burrow, and thus forms a closed cell, in which it is transformed into the pupa shown at Fig. 1a, remaining in this
condition until the warmer weather calls the insect from its
retreat. Two closely allied species are *T. perblandus* and
Metablax acutipennis. The former is occasionally found
under the large scales on matai trees, and resembles the
present insect in general appearance, but is much smaller and
more elongate in form, its elytra being also ornamented with
longitudinal rows of yellowish-brown hairs. The latter
may be often taken on the wing in the hottest sunshine,
and is chiefly remarkable for its elongate prothorax and
pointed elytra; its colour is dark reddish-brown, ornamented
with a few scattered white hairs. All these insects possess
the singular habit of leaping into the air when placed on
their backs, the last-named species exercising this faculty
in a most marked degree. The movement is effected by
the joint between the pro- and meso-thorax, the sternum
of the former being elongated into a long process, fitting
into a corresponding cavity in the latter, so that by means
of the two being suddenly brought together, the insect is
thrown high into the air with a loud clicking sound, hence
the English name of the Skipjack or Click Beetles, the
scientific name, Elater, doubtless having reference to the
same habit. The object of this curious arrangement is in all
probability twofold ; the sharp click and rapid movement
of the insect deterring many enemies from attacking it,
whilst the short legs of the beetle, which are quite unable
to reach the ground when it is thrown on its back, render
a special contrivance necessary.

Group HETEROMERA.

Family TENEBRIONIDÆ.

Uloma tenebrionides (Plate II., fig. 2, 2a larva, 2b pupa).

One of our commonest beetles, found in great abundance

in all moist wood when much decayed, the favourite trees
being apparently rimu and matai.   Its cylindrical larva
may be taken in similar situations, and much resembles
in general appearance the well-known "wire-worm" of
England, whose destructive habits, however, it does not
share.   At present, whilst bush-clearing is going on, its
influence is beneficial, as it devours large quantities of use-
less wood, which is thus rapidly broken up and got rid
of.   The pupa is enclosed in an oval cell, constructed by
the larva before changing, from which the perfect insect
emerges in due course.   When first exuded its colour is
pale red, but this rapidly changes into dark brown after
the insect has been hardened by exposure to the air.
Specimens are often met with of every intermediate
shade, and are rather liable to deceive the beginner, who
mistakes them for distinct species.   An account of a
small Dipterous insect infesting this beetle in its preparatory states will be found on page 62.

Group LONGICORNIA.

Family PRIONIDÆ.

Prionus reticularis (Plate II., fig. 3, 3b larva, 3a pupa).

This is the largest species of beetle found in New
Zealand, and is common throughout the summer in the
neighbourhood of forests.   Its larva (Fig. 3b) is a large,
fat grub, with minute legs; it inhabits rimu and matai,
logs, often committing great ravages on sound timber
although frequently eating that which is decayed; posts,
rails, and the rafters of houses alike suffer from its attacks;
the great holes formed by a full-grown larva of this insect
creating rapid destruction in the largest timbers.   It may
be remarked, in connection with these wood-boring species,
that a good thick coat of paint put on the timber as soon
as it is exposed, and renewed at frequent intervals, to a
great extent prevents their attacks.   The pupa (Fig. 3a)

is enclosed in one of the burrows formed by the larva, which, before changing, blocks up any aperture, so as to rest secure from all enemies. The perfect insect emerges in the following summer, when it may be often observed flying about at night. It is greatly attracted by light, and this propensity frequently leads it on summer evenings to invade ladies' drawing-rooms, when its sudden and noisy arrival is apt to cause much needless consternation amongst the inmates.

Closely allied to the above is *Ochrocydus huttoni*, which may be at once known by its smaller size and plain elytra; it is very much scarcer than *P. reticularis*, but may occasionally be cut out of dead manuka trees in company with its larva.

Group RHYNCOPHORA.
Family CURCULIONIDÆ.
Oreda notata (Plate II., fig. 4, 4a larva).

This weevil is not often noticed in the open, but may be found in great abundance in the dead stems of fuchsia, mahoe, and other soft-wooded shrubs, whose trunks are frequently noticed pierced with numerous cylindrical holes. The larva also inhabits these burrows, devouring large quantities of the wood; it is provided with a large head and powerful pair of mandibles, but, in common with all other weevil larvæ, does not possess legs of any description, the insect being absolutely helpless when removed from its home in the wood. The pupa might also be found in similar situations, but I have not yet observed it. The perfect insect may be cut out of the trees throughout the year, and is occasionally taken amongst herbage during the summer.

Family CURCULIONIDÆ.
Psepholax coronatus (Plate II., fig. 5 ♀, 5a ♂).

This curious species is found abundantly in the stems of

dead currant trees (*Aristotelia racemosa*), in which it excavates numerous cylindrical burrows like the last species,
which it closely resembles when in the larval state. The sexes
are widely different, the elytra of the male being furnished
with the characteristic coronet of spines, which is entirely
wanting in the female. Numerous other members of this
genus may be taken in company with the present insect,
and should be carefully examined, as a correct determina-
tion of the males and females of the several species is
sadly wanted. Digging beetles out of the wood is good
employment for the entomologist in winter, when he will
find that a day spent in this manner will frequently produce
as rich a harvest as one in the height of summer.

Before finally leaving the Coleoptera, I should like to
direct the attention of my readers to the immense number
of interesting weevils found in New Zealand. Chief
among these is the remarkable *Lasiorhynchus barbicornis*,
a large insect furnished with a gigantic rostrum, which will
at once distinguish it from any of the rest. Other genera
contain numerous beetles, which may be found in various
kinds of dead timber in company with their larvæ, and are
worthy of a more minute investigation than has at present
been given them.

CHAPTER IV.

The Hymenoptera.

THE Hymenoptera are perhaps the most interesting order of insects, their brilliant colours, great activity, and unparalleled instincts rendering them alike attractive to the young collector and scientific entomologist. They are, however, not very numerous in New Zealand, several of the most important families being completely absent ; in fact, with the exception of the ants, there are no social Hymenoptera native to this country. The information I here give in connection with these insects does not adequately represent the large amount of interest which can be derived from their investigation, and I must therefore refer the reader to those admirable works by Sir J. Lubbock on Ants and by Huber on Bees, which cannot fail to interest all who read them.

Family ANDRENIDÆ.

Dasycolletes hirtipes (?) (Plate III., fig. 1).

This is the true native bee of New Zealand, and may be taken abundantly during the whole of the summer. Its nest is constructed in crevices in the bark of trees, &c., the insect very frequently selecting the spaces between the boards of outhouses, where the loud buzzing noise

4

made by the perfect bees when emerging from their
retreat at once arrests our attention. These nests consist of about ten oval cells, formed of clay, and neatly
smoothed within. They are all constructed by a single
female, which also provisions them with honey and pollen,
depositing an egg in each   The larva, after consuming
the food, changes into a pupa, from which the perfect
insect emerges about January. If the reader will imagine
a great number of these nests closely packed together,
the formation and storing of the cells being performed
by a number of sterile individuals (workers), while the
eggs are deposited by a single female (queen), he will
have a fair idea of the economy of the social bees and
wasps, whose wonderful instincts attain their maximum
in the well-known hive-bee, successfully introduced and
cultivated in various parts of the country.

Closely allied to this species is *Dasycolletes purpureus* (?)
(Fig. 10), which forms its nests in sand-banks, its cylindrical holes having a great resemblance to the burrows of
Cincindela tuberculata, which frequently occur in the same
situation.

Family SPHEGIDÆ.

Pompilus fugax (Plate III., fig. 2).

This is a very abundant insect, and may be observed
flying about on any fine day during the summer, occasionally stopping to examine leaves and crevices in the
bark of trees, where it is looking for the unfortunate
spiders, which constitute the food of its progeny. The
larva is a fat apodal grub, and may be found in the cells
constructed by the perfect insect, which usually selects
a large cylindrical hole in a log, previously drilled out
by a weevil. Into this burrow she pushes a large quantity of spiders, which she has previously captured and
paralyzed with her venomous sting. When her nest is

properly provisioned she deposits an egg in it, closes the hole with a neat plug of clay, and leaves the larva to quietly consume its half-dead companions. Each female, no doubt, forms a large number of these cells during the summer. While cutting up old logs for Coleoptera, the entomologist will not infrequently come across these nests, when the insects may be found in various stages of development. Unfortunately, however, the sight which usually meets his eye is a large number of legs and other fragments of spiders, the *fugax* having long since deserted the burrow, and being very probably engaged in forming others in a neighbouring tree. These insects are very ferocious, and will attack spiders which considerably exceed them in size. On one occasion I noticed a very large one at rest in the centre of its web, which was suddenly noticed by a passing *fugax*, which immediately sprang upon its back, and, in spite of violent movements on the part of the spider, twisted her abdomen dexterously round and stung her victim in the centre of the thorax, between the insertions of the legs. This produced almost instantaneous paralysis in the spider ; but it was apparently too large for the *fugax* to carry away to her nest, as I saw the unfortunate creature hanging helplessly in its web some hours after the occurrence.

Family FORMICIDÆ.

Formica zealandica (Plate III., fig. 3 ♂, 3a ♀, 3b ☿, 3c, cocoon).

This is one of our commonest ants, and may be noticed under logs and stones throughout the year. The nest consists of a number of irregular cavities dug out by the workers either in the ground or in soft rotten wood. Its size varies considerably, but the societies of this species are not usually so extensive as those of *Atta antarctica*, an insect I shall have occasion to refer to presently. The larvæ are minute apodal grubs, which are dependent entirely on the workers for food. When full grown they spin an oval cocoon of white silk, in which they are converted into pupæ, and these the patient neuter ants may be observed carrying away with great anxiety when disturbed, risking their own lives to preserve their adopted offspring from destruction. The females, or queens, of which there are several in each nest, do not appear to participate in these labours, but are only instrumental in perpetuating the species, and the same remark applies to the males. A large number of these winged males and females may be observed in the nests about February, the general emergence taking place during that month. At this time they leave their native homes and mount to a great height in the air, and after sporting for some hours they re-alight on the earth, and in a short space of time cast their wings. The neuters at this time are said to carry them away to form fresh colonies, but I have not carried my investigations sufficiently far to verify this in connection with the New Zealand species.

Family FORMICIDÆ.

Ponera castanea (Plate III., fig. 4 ♂, 4a ☿, 4b, larva).

This is a much larger species of ant than the last, but is apparently not unlike it in habits. I have figured a male (Fig. 4) and worker (4a), the female not differing from the latter in any great degree, except in being provided with wings. It will be noticed, however, that the male is very divergent. The larvæ of this insect are covered with numerous minute spines, and may be often found in the nests; also the cocoons which they form when full grown, these latter being of a dark brown colour, and rather elongate. The winged insects are not frequently seen. They appear only for a short time in February, the earlier ones being invariably held captive by the workers until the rest have emerged, when they are all allowed to fly away and form fresh colonies as in the last species.

Family FORMICIDÆ.

Atta antarctica (Plate III., fig. 5 ♂, 5a ♀, 5b, larva).

This is another very abundant species, found occasionally amongst rotten wood in very large communities. Its larva, which is represented at Fig. 5b, does not form any cocoon, the pupa being quite naked and defenceless. It is a beautiful little object when examined with a microscope of moderate power. The annual migration of the winged males and females of this species usually takes place on a hot day in the last week of March, at which time I have observed the air throughout a day's journey absolutely swarming with these little insects. Many specimens are captured in the spiders' webs, while the logs, fences, and ground are covered with ants in the proportion of about ten males to one female. At other seasons of the year the winged individuals of *Atta antarctica* are seldom observed.

Family CHALCIDIDÆ.

Pteromalus, sp. (?) (Plate III., fig. 9).

This little insect was reared, in company with thirteen others of the same species, from a pupa of *Eurigaster marginatus*, which had been procured from a larva of *Œceticus omnivorus*, and is consequently a true hyperparasite.[1] Its curious habits will be better understood by the reader after perusal of the life-histories of those two insects, which I have given on pages 60 and 74. The method by which the females of the Hymenoptera whose larvæ are parasitic on insects inhabiting other insects,

---

[1] Hyperparasite is an animal parasitic in a parasite.

introduce their eggs into their hosts,[1] is not at present known to entomologists, but it seems at least probable that they are deposited in the eggs of the parasitic Dipteron before these gain access to the caterpillar of the moth.

Family ICHNEUMONIDÆ.

Ichneumon sollicitorius (Plate III., fig. 6).

This is the most abundant of our ichneumon-flies, and may be taken amongst herbage from August till May. Its larva is parasitic in the caterpillars of various Noctuæ, having occurred in the following species : *Mamestra composita*, *M. mutans*, and *M. ustistriga*. The pupa may be frequently discovered inside that of the moth, and is quite white in its early stages, but as age advances all the colours of the future insect can be seen through the thin pellicle which invests it. The perfect insect makes its escape through a circular hole, which it drills in the upper end of the unfortunate moth pupa it has destroyed. The sexes of all ichneumon-flies may be at once recognized by the females possessing an ovipositor[2] differing considerably in length among the various species, but nearly always plainly visible.

Family ICHNEUMONIDÆ.

Ichneumon deceptus (Plate III., fig. 7).

This conspicuous insect is chiefly mentioned on account of a very curious habit possessed by the females of congregating in large numbers on matai trees, as many as fifty or sixty specimens being often found huddled together under a single flake of the bark. The males are occasionally taken flying in the open, but I have never seen any amongst these large assemblages of females. Whether the

---

[1] " Host " is a term applied to any animal harbouring a parasite.
[2] Ovipositor, a boring instrument employed in depositing the eggs.

ichneumons are parasitic on some insect which lives on the matai, or whether they assemble to feast on the sweet juice occasionally exuded from its bark, it is impossible to say, but in either case the complete absence of males is a very remarkable circumstance.

Family ICHNEUMONIDÆ.

Scolobates varipes (Plate III., fig. 8).

The larva of this little insect is parasitic on the useful larva of *Syrphus ortas* whose life-history is recorded on page 57. It is very common in some instances, and must consequently destroy a considerable number. It entirely eats the soft portions of the insect, and may afterwards be found lying snugly within the hard empty shell of the deceased syrphus pupa, which acts as a cocoon for it while undergoing its own pupa state. The perfect insect may be often observed amongst herbage, searching for syrphus larvæ to deposit its eggs in.

CHAPTER V.

The Diptera.

THE next Order which comes under review is the Diptera, which includes all the two-winged insects, and constitutes a most extensive Order in respect to the number of distinct species. When, however, the numbers of individuals of the same species are considered, it is probable that this Order includes a greater proportion of the insect-world than all the others put together. The preponderance of these insects over the rest holds good with greater force in New Zealand than in many other countries, and this fact may be almost inferred from the large number of spiders present here, which are chiefly dependent on Diptera for their support. The important function of clearing away refuse matter is almost entirely performed by the members of this Order, as the Necrophagous Coleoptera and other scavengers which exist in such large numbers in many countries are practically absent here, and their work consequently devolves upon dipterous insects.

Group NEMOCERA.

Family CULICIDÆ.

Culex iracundus (Plate IV., fig. 1, 1a larva, 1b pupa).

The mosquito is only too familiar to every one from

its ceaseless attacks ; it occurs almost everywhere, but is
most abundant in marshy situations. The larva (Fig. 1a)
inhabits all stagnant waters, where it may be found very
abundantly throughout the summer, and when disturbed it
plunges about with great agility. Its food consists of the
numerous animalculæ swarming in all still waters during
the greater portion of the year. These are captured by
means of two curious anterior appendages, which are
fringed with long hair, and pulled through the water like a
fisherman's net ; they are then withdrawn into the mouth
and the contents devoured, the hungry insect again extending them for a fresh supply. These larvæ are generally
seen suspended from the surface of the water by the
curious air-tube which takes its rise from the penultimate
segment of the abdomen, which is of considerable length.
Its apex is armed with a row of stiff bristles, which
effectually prevent the water from entering the spiracle
there situated, so that the insect is enabled to respire when
hanging from the surface, independently of any muscular
action. It is also worthy of note that the intestine dis-
charges itself into this tube, an arrangement which does
not exist among the British species. After several moultings
the transformation to the pupa state takes place. At this
stage the insect (Fig. 1b) becomes much thickened anteriorly,
this being the region of the head and thorax of the future
gnat ; all the limbs are easily detected on a close examination, as with lepidopterous pupæ. The upper portion is
provided with two short appendages, fulfilling the same
function as the air-tube of the larva, and which constantly
support the pupa at the surface of the water. The terminal
fins enable it to dash through the water with great rapidity
when pursued by enemies ; at other times it remains perfectly motionless, suspended from the surface of the water.
It should be mentioned that none of these aquatic pupæ
take any nourishment, neither have they any limbs properly

so called. Their locomotion, although in some cases unquestionably rapid, is entirely effected by violent motions
of the abdomen. I have been careful to point out these
peculiarities as these animals have been regarded by
many authors as *active* pupæ on a level with those of the
Orthoptera and Hemiptera. This opinion, however, is
manifestly erroneous ; the pupæ of the nemocerous
Diptera are on precisely the same footing as those of the
Lepidoptera, and it would be almost as reasonable to call
one of these *active*, because it wriggles out of its cocoon in
the earth before the emergence of the moth. The perfect
mosquito emerges from a rent in the thoracic shield of the
pupa, drawing each pair of legs out separately, and placing
them in front of it on the water ; the wings and abdomen
are then extracted and in a few moments it flies away.

The bites of these insects appear to distress some
people much more than others, probably owing to constitu-
tional differences. I should mention that the females
alone engage in these attacks, the males being quite
harmless and subsisting entirely on honey, which is doubtless the natural food of both sexes. The male and female
mosquito are readily distinguished, the specimen figured
belonging to the latter sex ; her companion is chiefly
remarkable for his plumed antennæ and beautiful palpi,
which are very long and gracefully plumed. As many of
the harmless insects which will be investigated are often
mistaken for this species, and destroyed accordingly, I
should like to advise my readers that they may at once
distinguish all the venomous species of gnats by their long,
lancet-like proboscis and loud humming noise during flight.

Closely allied to this insect is *Culex argyropus*, which
might be called the coast mosquito as it is always found
near the seashore, its larva living in brackish pools
just above high-water mark. The perfect insect may be
also seen skating along the surface of the water like a

gerris [1] ; it may be at once distinguished by its dark colour.

Family TIPULIDÆ.

Corethra antarctica, n.s.[2] (Plate IV., fig. 3, 3a larva, 3b pupa).

An elegant little gnat, frequenting the margins of ponds and ditches during the spring months. The larva (Fig. 3a) is bright green, ornamented with numerous yellow spots ; it is very sluggish, living in the green slime weed which floats on the water in such large masses during that season. Not being very common it is difficult to find, as its colour so closely resembles that of the weed which it always frequents. The pupa (Fig 3b), is not very agile, and is nearly always observed suspended from the surface by its thoracic air-tubes and caudal fins, the abdomen being directed upwards and thus bringing the two pairs of organs close together. In its metamorphosis and general appearance this insect forms a convenient link between the present family and the Culicidæ.

Family TIPULIDÆ.

Chironomus zealandicus, n.s. (Plate IV., fig. 2, 2a larva, 2b pupa).

This is the common midge of New Zealand, and is extremely abundant throughout the country. Its larva (Fig. 2a) inhabits the soft mud at the bottom of stagnant ponds and streams, and is very conspicuous, being of a brilliant crimson colour and thus much resembling the well-known " Bloodworm " of English anglers, which is the larva of a closely allied European species (*C. plumosus*). It may be readily kept in an aquarium, and if supplied with a little soil and green weed will rapidly cover the

---

[1] A genus of Hemipterous insects commonly seen skipping over ponds in England.

[2] " n.s." is the accepted abbreviation for new species."

walls of its glass prison with numerous tubular galleries.
These take their rise from the mud at the bottom, and,
extending upwards to a distance of three or four inches,
afford the larva a convenient retreat from all enemies.
These insects are occasionally seen swimming laboriously
through the water with a peculiar zigzag motion. When
out of their burrows they have considerable difficulty in
keeping beneath the surface, and may be often observed
floating helplessly with their exposed portions quite dry ;
in fact the whole integment of the insect appears to have
a peculiar power of resisting the water. The pupa (Fig.
2b), is a most beautiful object, its anterior extremity being
obtusely thickened and the limbs of the future insect quite
discernible. On each side of the thorax the gills form a set
of graceful plumes, a much smaller group being also situated
at the extremity of its abdomen. In this state the insect
remains almost entirely concealed in the burrows previously
constructed by the larva, its gills imbibing sufficient air
from the surrounding medium, and thus rendering ascen-
sion to the surface unnecessary. The water is periodically
circulated in the tunnels by violent movements on the
part of the pupa. About a day before emergence the
insect assumes a peculiar silvery appearance, which is
occasioned by the presence of a large quantity of air
between the imago and its pupa skin. This air has been
first imbibed by the gills and afterwards expelled through
the spiracles of the enclosed gnat, thus inflating the skin of
the pupa, and helping to buoy it up during its last and
most important transformation. Leaving its tunnel the
insect rises to the surface, the thorax is lifted above the
water which retreats from it on all sides, the skin cracks
open at the back and the insect slowly extricates itself in
a similar manner to the mosquito. In about ten minutes'
time the wings are sufficiently hardened for use and the
insect then flies ashore, but we may occasionally notice,

beside their old pupa-skins, drowned individuals which have failed to effect a successful emergence. The perfect insect is extremely common in all swampy situations throughout the summer ; it has a great partiality for light, and may be occasionally noticed in vast numbers round the street lamps on a hot summer's night, especially if rain is impending. It is a most graceful insect, and will amply repay a minute examination (Fig. 2).

Family TIPULIDÆ.

Ceratopogon antipodum, n.s. (Plate IV., fig. 4, 4a larva, 4b pupa).

Very plentiful in the forest throughout the year, often enlivening the winter sunshine by its merry gambols. The larva (Fig. 4a), is found under the bark of newly fallen trees, feeding on the sap which exudes in large quantities from the logs whilst drying. When first discovered it often has a curiously spangled appearance, owing to the minute beads of moisture retained by numerous bristles clothing the larva. When about to change, these insects assemble in large companies of thirty or forty, firmly affixing their basal segments to the wood, their heads all pointing inwards and forming a small circle. In some cases, where an unusually large gathering has occurred, a number arrange themselves into an outer row, their heads being immediately behind the extremities of the inner group, the whole thus bearing a rough likeness to the radiations of a star-fish. The pupa is very short, and is furnished with two clubbed horns on the thorax for respiration. Its abdominal portions are retained within the old larval skin, thus keeping it firmly anchored to the log. The perfect insect emerges from a rent in the thorax of the pupa, groups of exuviæ being of common occurrence under the bark. The sexes differ considerably, the individual figured

(Fig. 4) being a male; the female is slightly larger, and
much more stoutly built; her antennæ are filiform[1], and
the limbs generally shorter. Both are equally common,
but the male is more often noticed, owing to his greater
activity.

Family TIPULIDÆ.

Psychoda conspicillata (Plate IV., fig. 6).

A common species, occurring plentifully on window
panes during August, and bearing a great superficial
resemblance to a small moth of the Tineina group, often
deceiving the novice in consequence. It is a beautiful
object for the microscope, the figure being a careful drawing of the insect, seen with a power of about ten diameters.
I regret to say that its transformations are at present unknown.

Family TIPULIDÆ.

Mycetophila antarctica, n.s. (Plate IV., fig. 5, 5a larva,
5b pupa).

Tolerably common in the vicinity of forest during the
major part of the year. The larva (Fig. 5a), is a small
elongate maggot of a pinkish colour; it is a social insect,
inhabiting rotten pine logs, which it perforates with numerous cylindrical burrows. These larvæ, entirely confine their
attention to damp wood of a "pappy" consistency, leaving
the harder logs for the wood-boring Coleoptera, which are
provided with much stronger jaws. They consequently
do not injure the rafters and boards of houses, or other
valuable timbers. The pupa (Fig. 5b) is very elongate,
reposing in one of the burrows, previously constructed by
the larva. It probably breathes by means of its spiracles,
as no special organs of respiration are visible. The perfect
insect appears in a short time, flying sluggishly in the sunshine, the female possessing an enormous abdomen, which

[1] Thread-like.

almost incapacitates her for aerial locomotion ; in other respects she resembles the male, which is the sex figured (Fig. 5).

Family TIPULIDÆ.

Tipula holochlora (Plate V., fig. 1, 1a larva, 1b pupa).

This beautiful insect is very common in the forest throughout New Zealand. Its larva (Fig. 1a) inhabits various kinds of decaying wood, frequently occurring in vegetable refuse at the roots of trees. It is a large, sluggish-looking grub, and the anterior segments are very retractile. Its colour appears to vary according to its surroundings, those specimens found in red pine being of the dull reddish hue characteristic of that wood, while those taken from pukatea and henau are dark brown larvæ, resembling the illustration. These insects are very voracious, but their growth is gradual, each larva probably occupying at least six months to reach maturity. They mostly feed during the winter, but may be often taken at other times. The pupa (Fig. 1b) is enclosed in a small oval cell, previously excavated by the larva, which also constructs a ready means of escape for the future insect in the form of a small tunnel leading out of one end of its prison to the open air. Through this the pupa wriggles, assisted by the spines, which arm the edges of all the segments ; the coronet of hooks at its extremity retaining the insect firmly at the mouth of its burrow while undergoing its final transformation. After numerous twistings and contortions on the part of the pupa, a rent is formed in the thoracic plates, and the imago draws itself out, standing on the log until its wings are sufficiently hardened for flight. In many old houses numbers of these exuviæ may be seen projecting from holes in the boards—a relic of the destruction that has taken place within. These insects naturally inhabit dead trees, but as they will devour unsound timber in any form they are very injurious to old wooden buildings. The perfect insect chiefly frequents forest, where it is difficult to detect owing to its green colour harmonizing so closely with the leaves. The specimen figured (Fig. 1) is a male, the female being considerably smaller with a much stouter body and shorter legs.

Family TIPULIDÆ.

Tipula fumipennis, n.s. (Plate V., fig. 2, 2a larva, 2b pupa).

Another fine species, occurring in similar situations to the last, but not quite so commonly. The larva (Fig. 2a) may be found throughout the year under the bark of very rotten henau and pukatea, feeding on the moist decaying wood. It constructs in this material numerous burrows, which are lined with a viscous fluid constantly emitted from the mouth. Its movements in these are very rapid, frequently eluding the most careful searches. When divested of its slimy covering, it is anything but an offensive-looking larva, the great air-tubes, which run the whole length of the insect, being very conspicuous, and many of the other internal organs are easily detected owing to its partial transparency. The pupa (Fig. 2b) is enclosed in a small cocoon, having ready access to the air ; it is chiefly remarkable for its very large thoracic horns, which are curiously toothed. The air-tubes connected with these are distinctly visible in the abdomen of the insect, where they may be seen branching in all directions. When about to emerge this pupa works its way to the surface of the log, the head and thorax are thrust outside, and the perfect insect escapes in the ordinary way. The illustration (Fig. 2) is taken from a female ; the male differs in being less robust, and in being provided with longer legs.

Family TIPULIDÆ.
The Glow-worm. *Bolitophila luminosa*, Skuse.
(Frontispiece, fig. 1).

Every one who has walked in the forest at night has no doubt noticed, in many damp and precipitous situations, numerous brilliant points of greenish white light shining out from amongst the dense undergrowth. The animal which causes this light may be seen at Fig. 1a on the Frontispiece, and is probably one of the most interesting insects we have in New Zealand. It inhabits irregular cavities, mostly situated in the banks of streams, where it hangs suspended in a glutinous web which is stretched across the cavity and supported by several smaller threads running right and left, and attached to the sides and ends of the niche. On this the larva invariably rests, but when disturbed immediately glides back along the main thread and retreats into a hole which it has provided at the end of it. From the lower side of this central thread numerous smaller threads hang down, and are always covered with little globules of water, constituting a conspicuous, though apparently unimportant, portion of the insect's web. It should be mentioned that all these threads are constructed by the larva from a sticky mucus exuded from the mouth.

The organ which emits the light can easily be seen by referring to Fig. 1a. It is situated at the posterior extremity of the larva, and is a gelatinous and semi-transparent structure capable of a great diversity of form. It can be extended or withdrawn at the will of the larva, which, however, can shut off the light independently of this latter action. Larvæ cease to shine on very cold nights, in the daytime, and in a room which is artificially lighted. They gleam most brilliantly on dark, damp nights, with a light north-west wind. These larvæ appear to suffer great mortality in a state of nature, as the young ones will always be found greatly in excess of those that are approaching maturity.

When full-grown this insect is transformed into the curious pupa shown at Fig. 1b. It is furnished with a large process on the back of the thorax which is attached to the web and holds the pupa suspended in the middle of the niche previously inhabited by the larva. The light is emitted from the posterior segment of the pupa, but is much fainter than in the larva, and a distinct organ is not apparent. It is frequently suppressed for days together.

The perfect insect is drawn at Fig. 1. It emits a strong light from the posterior segment of the abdomen, about half as bright as that emanating from a full grown larva. It has been recently described by Mr. Skuse, of Sydney, as *Bolitophila luminosa.*

During the whole course of my observations[1] on this insect, extending over five years, I have only succeeded in bringing two specimens to maturity, and both of these were females.

The uses of the light and the web to the larva are at present quite unknown to me, as well as its food, which, however, possibly consists of fungi. It should also be mentioned that the larvæ are found in the greatest abundance in mining tunnels, many feet below the surface of the earth, as well as in caves.

Family TIPULIDÆ.

Cloniophora subfasciata (Plate V., fig. 3, 3a larva).

Tolerably common in damp gullies during summer and autumn. The larva (Fig. 3a) inhabits decayed henau logs,

[1] For an extended account of these observations see " Transactions of the New Zealand Institute," vol. xxiii. (1890).

drilling deep into the wood, where its burrows are seldom noticed, as they are filled up with refuse almost as soon as they are made. The pupa resembles that of *Tipula holo-chlora*, but is rather more attenuated in the body, and the thoracic horns are slightly thicker. It is not enclosed in any cocoon, but lies amongst the powdery wood, wriggling to the surface when about to emerge. The illustration represents the male insect, the female having a much stouter body, with short thick legs ; she also differs in her antennæ, which are much less branched than those of the male.

Family TIPULIDÆ.

Rhyphus neozealandicus (Plate V., fig. 4, 4a larva, 4b pupa).

A most abundant species occurring in most damp situations throughout the year. Its larva (Fig. 4a) closely resembles a small worm, being of an elongate form attenuated at each end. The skin is very hard and of a dull yellow colour, with black markings. The food of this insect consists of decaying vegetable matter, which it procures by means of two small appendages, situated on each side of the mouth, and which it is continually moving about in search of suitable materials. The pupa is a curious object (Fig. 4b), the two little respiratory horns having a singular resemblance to a pair of ears. It is enclosed in a small oval cell about one inch below the surface of the earth, the insect working its way to the air before emergence. The perfect *Rhyphus* may be almost regarded as one of our domestic insects, and is seldom found in the open country, but frequents cowhouses and other farm buildings in great numbers, the larvæ feeding on the manure in these situations. It is often mistaken by ignorant people for the mosquito and at once destroyed, but quite unfairly, as the species is in reality perfectly harmless, frequently

benefiting mankind by the removal of considerable quantities of effete matter, which if allowed to remain could
not fail to be injurious.

Family TIPULIDÆ.

Bibio nigrostigma (Plate V., fig. 5, 5a larva, 5b pupa).

This insect is very abundant during the spring months,.
but rapidly disappears, and few specimens are noticed after
Christmas.   Its larva (Fig. 5a) inhabits the woody powder
often found under logs, which frequently consists of the
accumulated excrement of wood-boring insects.   It is gre-
garious in its habits, being found in large companies of fifty
or a hundred individuals.   When first disturbed these appear as a wriggling mass, but very shortly become so still
that they can only be distinguished with the greatest difficulty from morsels of bark.   A considerable portion of the
powdered wood is also retained on the body of the insect
by a row of short spines situated in the middle of each
segment, which helps to render the larva still more inconspicuous.   In this condition it remains for at least eight
months, during which time growth takes place very slowly.
About September the larvæ separate, each being afterwards
transformed into a small yellowish pupa (5b), whose abdominal extremity is usually retained within the old skin,.
thus closely resembling that of the genus *Ceratopogon.*   I
have figured this pupa entirely naked, in order to show
its characteristics, some of which are rather remarkable,.
more completely, the agglutination of nearly all the
anterior portions of the body being especially noteworthy.
The perfect insects may be found everywhere, the males
sucking honey from the flowers and performing many
antics in the air, often clinging hold of one another and
whirling about together.   The female seldom flies, but is
usually observed crawling about fences or the trunks of
trees.   She may be at once recognized by her heavy body

which is very large when distended with eggs. Her general colour is dull red, thus differing widely from the male insect represented in the illustration (Fig. 5).

Family TIPULIDÆ.

Simulia australiensis (Plate VI., fig. 1, 1a larva, 1b pupa).

Every one knows the sandfly, the little black insect that so persistently perches on our hands and faces and inflicts its painful punctures, which in many cases are followed by large swellings, often lasting for several days and causing much irritation. Its larva (Fig. 1a) inhabits clear running water, climbing about in strong currents by means of a pair of suckers situated at each end of the body, two being placed on the prothoracic segment just behind the head and two others close to the anal extremity. These the insect employs rather curiously, the anterior pair being first affixed and the others drawn up close behind them, its elongate body consequently forming a loop. Clinging by the posterior suckers for a moment the larva then reaches forward, re-affixes the anterior ones, and draws up the posterior as before. Breathing is performed by two spiracles situated on the last abdominal segments near the hind pair of suckers. Two large air-tubes originate from these and run forwards, giving off branches to all parts of the body ; they terminate in a number of air-sacs in the thorax. The food of this larva consists of animalculæ, which are no doubt obtained by drawing the two ciliated appendages rapidly through the water several times in succession, their contents being afterwards gathered up by the smaller organs and passed into the mouth. When about to assume the pupa state the insect covers itself with a glutinous envelope, which is firmly joined to the under side of a leaf, the transformation taking place within a few days. The pupa can hardly be distinguished from a small moth chrysalis except for a pair of branching fila-

ments, which arise from the top of the thorax and serve
the purpose of gills (Fig. 1b).   Before emergence the
anterior segments are projected nearly out of the cocoon
from which the perfect sandfly makes its escape, and floating to the surface of the water ascends the stem of an
aquatic plant to expand its wings.   I should here remark
that as with the mosquitoes, the bloodthirsty propensities of
the present species have no doubt been acquired since the
arrival of man and other warm-blooded animals.

Group BRACHOCERA.

Family TABANIDÆ.

Tabanus impar (Plate VI., fig. 6).

I have figured this fine species as a representative of a
most important family of Dipterous insects, but am at
present quite unacquainted with its life-history.   It occurs
plentifully on the margins of the forest throughout the
summer.

Family BOMBYLIDÆ.

Comptosia bicolor (Plate VI., fig. 2).

This conspicuous species is very abundant in glades
throughout the summer, flying with great rapidity, and
delighting to suck honey from the numerous shrubs which
are in blossom at that time of year.   It is a social species,
and is usually found in companies of fifteen or twenty
individuals, which engage in endless dances, two insects
often seizing one another on the wing and then re-
volving together like a wheel in rapid motion.   Their
manœuvres in avoiding the strong gusty wind, so often
prevalent in early summer, are also interesting ; the insects
play upon the wing whilst the air is quiet, but if a breeze
springs up they instantly settle on the nearest bush, rising
to renew their sports when it is again calm.   These flies are
rather variable in colour, some specimens being dark brown,

whilst others are more or less covered with greyish-white hairs; individuals are also often met with quite black and shining, their hirsute covering having been completely rubbed off. The female may be at once recognized by her solid, fleshy abdomen, that of the male being inflated by two great air-bladders, which cause that portion of the body to appear semi-transparent when the insect is held up to the light. The figure (2) is taken from a specimen of the latter sex.

Closely allied to the present insect is *Comptosia virida,* n.s. (Fig. 3), which can be at once distinguished by its brilliant green eyes and pale grey clothing. The larva of this species is a large white maggot, rather robust, and possessing a small head. It inhabits the dense moss growing on the trunks of trees in the forest, feeding on the roots of these plants, and finally forming an oval cocoon, in which it changes into the pupa shown at Fig. 3b. The perfect insect appears in a few weeks' time, when it may be taken in similar situations to *C. bicolor,* but in much fewer numbers.

Family ASILIDÆ.

Sarapogon viduus (Plate VI., fig. 4, 4a larva, 4b pupa).

A voracious insect, frequenting all dry sand-banks and pathways throughout the summer, and destroying the numerous minute diptera found in those situations. These unfortunate victims are drilled through the thorax by their destroyer, which sucks them completely dry with its long beak-like proboscis. The larva (Fig. 4a) inhabits rotten wood, chiefly feeding upon the moist, powdery portions. It is usually somewhat sluggish, but when disturbed hops about with electrical rapidity. The head is very minute, and the elongate body consists of twenty segments, a number very unusual among larvæ, the normal number being twelve exclusive of the head. It lives for a considerable time and is finally transformed into the blunt-looking pupa, drawn at Fig. 4b, without having previously constructed any cocoon. From this the perfect insect emerges in a month or six weeks' time, commencing its work of destruction as soon as its wings are hardened, which takes place within a few hours.

Family STRATIOMIDÆ.

Exaireta spiniger (Plate VI., fig. 5).

Abundant during November, when it may be taken in great numbers in the vicinity of water. The larva is probably aquatic, but I have not yet observed it, although its habits would, no doubt, be very interesting. The perfect insects frequent flowers, and are generally very sluggish in their movements.

Family ACROCERIDÆ.

Acrocera longirostris, n.s. (Plate VII., fig. 4).

An extraordinary and very rare species, occurring amongst white rata[1] blossoms in February. At present I have only taken three specimens, viz., two in Wellington and one in Nelson. The transformations of all the Acroceridæ are as yet unknown.

Family SYRPHIDÆ.

Syrphus ortas (Plate VII., fig. 3, 3a larva, 3b pupa).

Very common everywhere from September till May, or even later, when specimens may be often seen basking in the winter sunshine. The larva (3a) is a most useful insect to gardeners as it destroys an immense number of aphides, those noxious little insects that commit such fearful ravages on many valuable plants (see Hemiptera, page

[1] Metrosideros scandens.

120). In general appearance this larva resembles a small green slug, with the skin much wrinkled, and bearing at its extremity a short thick tube, which is probably the respiratory apparatus, the four lunate holes situated at its apex being no doubt the spiracles. These insects grow very slowly, occupying several weeks to attain maturity. Their mode of capturing the aphides is very curious, and is, briefly, as follows :—The larva lies in the midst of a number of aphides, and it occasionally happens that some of them crawl over it. On feeling an aphis touch its back the larva instantly darts out its long, pointed head and strikes its prey with the apex, which is enveloped in a quantity of very sticky mucus constantly ejected from the mouth. On the aphis being thus captured the larva withdraws its head into the hinder segments of its body and devours all the juicy portions of the aphis, whose dry skin is afterwards thrown aside. When full-grown it slowly shrinks up and changes into the pupa shown at Fig. 3b. In this state it is not protected by any kind of cocoon, but lies amongst the refuse of the aphides, near the stem of the plant. The fly emerges in a fortnight or three weeks' time, and is very fond of hovering over and sucking honey from the flowers, but the females may be often noticed running about plants, probably in search of a suitable place to oviposit.[1] For an account of *Scolobates varipes*, a species parasitic on the present insect, I refer to page 39.

Family SYRPHIDÆ.

Eristalis cingulatus (Plate VII., fig. 2).

This conspicuous insect occurs occasionally in glades in the forest about January, but is by no means common. It is very fond of the white rata flowers, where it may be

---

[1] Or lay eggs.

taken, if anywhere. Its life-history is at present unknown, but no doubt resembles that of the following insect.

Family SYRPHIDÆ.

Helophilus trilineatus (Plate VII., fig. 1, 1a larva, 1b pupa).

This fine species occurs abundantly in all damp situations throughout the summer. Its larva may be found in stagnant pools and is often met with in the mud at the bottom of ditches. Its posterior segments are enormously elongated, forming a telescopic breathing apparatus, composed of two tubes, the smaller of which is capable of being more or less extended at the will of the larva, which is thus enabled to adjust the length of its breathing tube, according to the depth of water or mud in which it happens to reside. This peculiarity has given all these larvæ the name of rat-tailed maggots. The other segments are very stout, each being furnished with a pair of minute feet, and the head is also provided with two small appendages which are supposed to be the outlets through which the exhausted air is discharged by the larva. When mature this insect leaves the water, forming a small oval cell in the neighbouring moist earth, in which it lies with its long tail folded along the breast. The skin then gradually hardens, and it is finally transformed into the pupa shown at Fig. 1b, the conical pair of breathing-tubes on the thorax being slowly protruded from two hardly perceptible warts, whilst the telescopic apparatus shrinks up, its functions being at an end. A variable time, dependent upon the season, elapses before the perfect insect makes its appearance, but prior to this occurring, a large circular plate, forming the thorax of the pupa, is thrust off, thus assisting the escape of the fly, which immediately ascends a plant, or other convenient object, to dry and expand its wings (Fig. 1). In the perfect state it delights to hover in the air, darting away with great rapidity on the approach of any enemies. It also frequently enters houses, where its presence is at once betrayed by a peculiarly shrill noise made while flying. The sexes of this insect differ chiefly in size, the female (Fig. 1) being about twice as large as her companion.

Closely allied to this species are *Helophilus ineptus,* and *H. hochstetteri.* The former is slightly smaller than *H. trilineatus* and may be at once distinguished by its tessellated orange-yellow and black abdomen. It is rather local, but extremely abundant wherever found. The latter has a superficial resemblance to some of the smaller blowflies (*Musca*), but may be readily known by its large brownish-red scutellum.[1] It is the commonest of the genus and may be found in great numbers throughout the summer amongst veronica and other flowers.

Family MUSCIDÆ.

Miltogramma mestor (?) (Plate VII., fig. 5).

A conspicuous species, found occasionally on forest-clad hills round Wellington. The life-history is at present unknown, but its larva is very possibly parasitic in some large Lepidoptera.

Family MUSCIDÆ.

Nemorea nyctemerianus (Plate VII., fig. 6).

This little fly is seldom met with in the perfect state. Its larva is parasitic on the caterpillar of *Nyctemera annulata,*[2] the eggs being deposited on the moth larva at an early age. The caterpillar grows and eats in the ordinary way, until it has assumed the chrysalis state, when the

[1] Scutellum : A horny plate situated on the mesonotum, usually somewhat triangular in form.
[2] For life-history of this insect see page 73.

maggot eats its way out and changes into a dark-brown pupa. In this condition the parasite is protected by the web which was previously constructed by the unfortunate caterpillar for its own use. The perfect fly appears in about six weeks' time, its great agility and large white scales rendering it very conspicuous.

Family MUSCIDÆ.

Eurigaster marginatus (Plate VII., fig. 7).

Another parasitic species, its larva inhabiting the caterpillars of various noctuæ which it destroys just before they change into the chrysalis state. The pupa of the parasite lies in a small oval cell constructed in the earth by its larva. A variable number of these maggots are found associated in one host, the smaller caterpillars only harbouring a single individual, while a large larva will frequently contain three or four. This species has been bred from the following Lepidoptera: *Mamestra composita*, *M. ustistriga* and *M. mutans.* It also occurs in the curious *Œceticus omnivorus*, being found in the cocoons of that moth in numbers varying from two to eleven, or even more, and it is especially interesting, as it is in turn destroyed by a small species of *Pteromalus* already noticed among the Hymenoptera (page 37). The perfect insect occurs occasionally on flowers throughout the summer.

Family MUSCIDÆ.

Calliphora quadrimaculata (Plate VII., fig. 9).

This is the large blue-bottle fly of New Zealand and is found everywhere in great abundance. Its larva feeds on decaying flesh and is of a dirty yellow colour, measuring, when full-grown, about seven lines in length. The pupa is buried at a considerable depth in the ground, the

larva having descended before changing. The duration
of this, and in fact of all the stages of the insect, depends
entirely upon the temperature, but the females invariably
deposit eggs, even during the hottest weather, and are
never ovo-viviparous like the next species, and several
others of the genus.

Family MUSCIDÆ.

Sarcophaga læmica (Plate VII., fig. 10).

Another extremely abundant species having a similar
history to the last, but its powers of development are
very much accelerated owing to the larva being positively
born alive. The females hover over meat and other
suitable substances, depositing a number of minute wriggling maggots thereon, not infrequently to the great
disgust of some hungry individual, who perhaps is making
his dinner off a mutton chop which the fly has selected
as a home for her offspring. These larvæ are all produced from distinct ova, which hatch before being laid,
as I have often proved, by removing them from the
insect's abdomen, and watching the young larva emerge
from a minute elliptical white egg, covered with a thin
leathery skin. Every one who has travelled in New
Zealand must have noticed that, in the wildest spots,
these insects assemble in large numbers as soon as any
meat is uncovered, thus not only showing their universal
distribution throughout the country, but also that they
possess a very keen sense of smell.

Two British species at least, allied to this genus, have
been introduced into New Zealand, *viz.*, *Musca domestica*
and *Musca cæsar*. The former is probably a world-wide
insect, every ship teeming with it, but the latter is
at present rather scarce and is usually found in the
neighbourhood of farm-yards, where the larva feeds on

cow-dung. The perfect insect may be at once known by
its brilliant green colour.

Family MUSCIDÆ.

Cylindria sigma (Plate VII., fig. 14).

A curious species, occurring occasionally in damp
situations in the forest where it may be noticed leisurely
walking over the leaves of various shrubs. It is very
sluggish and may often be captured between the fingers
without the aid of a net. Its life-history is at present
unknown, but the larva probably feeds on fungi. The
pretty little insect depicted at Fig. 11 may be found in
similar situations but is not so common.

Family MUSCIDÆ.

Phora omnivora, n.s. (Plate VII., fig. 15, 15a pupa).

This minute species may be found in large numbers
nearly all the year round. Its larva is parasitic on a great
variety of insects and is also not infrequently met with
among decaying vegetable matter. Its habits are, therefore, very varied. When parasitic in the Lepidoptera
it usually selects the noctuæ, destroying a great number
of many of the commoner species[1]. The infected caterpillars usually turn into chrysalides some time before the
little maggots emerge, but this is not invariably the case,
the parasite often destroying the larva at a comparatively
early stage. The pupæ are buried in the earth, near the
remains of their host, and are light brown in colour, with
the segments much more distinct than is usual (Fig. 15a).
From these the perfect flies proceed in about a month's
time. The occurrence of this insect as a parasite in
Coleoptera is not common, but I know of one instance

[1] Mamestra composita, M. mutans, M. ustistriga, Erana graminosa, &c.

in which a number of these little flies were produced from a pupa of *Uloma tenebrionides* (Plate II., Figs. 2, 2a, 2b), which I was rearing at the time (page 29). In this case it is difficult to understand how the female contrives to deposit her eggs in a horny beetle larva which lies safely hidden in its narrow tunnel in the middle of a large log of wood. Among bees this is a most destructive insect, its larva being parasitic in their grubs, and thus greatly reducing the population of the hive, which is finally ruined by the wholesale destruction of its honey when the flies emerge. Driving the bees into a fresh box would, no doubt, be frequently beneficial in these cases, but it is to be feared that bee-keepers will have much difficulty in contending with this insect. Its sexes are readily distinguished by their size, the female being considerably the larger.

Family MUSCIDÆ.

Coelopa littoralis (Plate VII., fig. 13).

Extremely abundant on the sea-beach. Its larva feeds on decaying seaweed, burying itself in the sand before changing. The perfect insects often congregate in such vast numbers on some of the rocks that it is necessary to run past them in order to avoid being positively suffocated by the countless multitudes which fly up into one's face. This insect must be regarded as the New Zealand representative of the well-known dungfly of England (*S. stercoraria*), which many of my readers will recollect has a similar habit of assembling in great numbers.

Family ŒSTRIDÆ.

Œstrus perplexus, n.s. (Plate VII., fig. 12).

This species is mentioned here as it is the only New Zealand exponent of a very important and well-known family of Dipterous insects. I am at present quite ignorant as to its life-history which would, no doubt, be very interesting. The only two specimens I possess were taken at Nelson, some four years back, so that it appears to be very rare.

The two remaining groups of the Diptera are of very limited extent. The *Pupipara* include a few anomalous species, in which the young are not deposited until they become pupæ, thus undergoing all their transformations within the body of the parent, while the *Pulicina* comprise the well-known fleas, which are probably identical with the European species. They are placed by many authors in a distinct order termed the *Aphaniptera.*

CHAPTER VI.

The Lepidoptera.

THIS Order includes the well-known Butterflies and Moths which are the first insects to arrest attention on account of their beautiful colouring and conspicuous appearance. Some of the families are fairly numerous in New Zealand, but the diurnal section is decidedly poorly represented, our total number of butterflies being limited to fifteen, of which one (*Diadema nerina*) has unquestionably been introduced from Australia, although it will doubtless shortly effect a permanent settlement in the Nelson district, where several specimens have recently been observed. Among the others only four species can be called at all common, the remaining twelve only occurring in certain favoured localities. Of the moths there are a large number, chiefly belonging to the Geometridæ and Micro-Lepidoptera, many of which are very interesting. Of the life-histories of the latter, however, I regret to say there is little known at present, the attention of naturalists having been hitherto chiefly occupied with the larger and more conspicuous species.

Group RHOPALOCERA.

Family NYMPHALIDÆ.

Argyrophenga antipodum (Plate VIII., fig. 1 type, 1a var.).

Passing over the local but conspicuous *Danais plexippus*,

about which so much doubt exists as to its origin in this
country, we come to *A. antipodum*, one of the most curious
and interesting butterflies found in New Zealand. It occurs
in great abundance amongst the tussock grass on the
plains in the South Island, but becomes an alpine species
further north. I have taken a very peculiar form (Fig.
1a) on the "Mineral Belt" near Nelson, but can find no
record of its appearance in the North Island at present.
Its larva is as yet unknown, but in all probability it feeds on
tussock grass, a fractured pupa having been found attached
to that plant by Mr. G. F. Mathew in January, 1884. Two
other closely allied species are *Erebia pluto* and *Erebia
butleri*, both strictly alpine insects, occurring in the South
Island at elevations ranging from 4,000 to 6,000 feet.

Family NYMPHALIDÆ.

Vanessa gonerilla[1] (Plate VIII., fig. 2, 2a underside,
2b 2c larvæ, 2d 2e pupæ).

One of our most beautiful butterflies, found abundantly
throughout the country from August till May. The
larva feeds on the New Zealand nettle, where it may be
taken in great plenty by careful searching. The caterpillar
joins several of the leaves together and forms a sort of tent,
in which it lives secure from all enemies. While young,
these insects are of a uniform dull brown colour, with two
faint lines on each side, but as age advances they become
very variable. The two extreme forms of variation are
depicted at Figs. 2b and 2c, the dark-coloured variety being
by far the commoner. When full-grown, this larva suspends itself by the tail to a small patch of silk, which it has
previously spun on the under side of a leaf. In this position it remains for about twenty hours, when it begins to
twist and distend the lower portions of its body, thus

[1] This genus, as represented in New Zealand, is often called
Pyrameis.

causing the skin to eventually break on the back of the thoracic segments, when the soft green pupa may be seen through the rent. The insect now works the skin upwards by violent wriggling motions until it is gathered in a crumpled mass round its tail, the old rent extending on one side almost up to the silken pad to which it is suspended. Through this rent the tail of the pupa is brought and firmly anchored in the silk by a few vigorous strokes, the insect hanging meanwhile to the skin which has not been quite cast off on the reverse side to the rent. When thus firmly attached to the silken pad, the pupa shakes itself entirely free, whirling itself round and round until the old skin is dislodged from the silk and falls to the ground. The two usual varieties of pupæ are shown at Figs. 2d and 2e, many of them being more or less ornamented with metallic gold or silver spots. The butterfly emerges in a fortnight or three weeks, and is common from February till April in most situations, but the greatest numbers are to be found in the spring months. These hybernated specimens appear as early as August, and some of them survive till the end of December or beginning of January, when the earliest of the new ones are just emerging. In fact it is not infrequent at this time to take both hybernated and recent specimens together. This species is a great traveller, and may be often seen flying over the tops of the trees at a great rate. It shows a singular indifference to shadow, and is constantly flying out of the sunlight into shady places in the forest, probably in search of the food-plant of the larvæ. The two other species of *Vanessa* are *V. cardui*, a periodical insect only distinguished from the "Painted Lady Butterfly" of England by the blue centres in three of the black spots on its hind-wings, and *V. Itea*, a lovely butterfly found in the northern portions of this island, of which I have at present only taken three specimens.

Family LYCÆNIDÆ.

Chrysophanus salustius (Plate VIII., fig. 3 ♂, 3a ♀, 3b larva).

This is the commonest of our Butterflies, and is found in great abundance throughout both islands from November till April. It is double brooded, and is consequently most abundant in the early summer and in the autumn, few of these merry little insects being seen at midsummer. The most forward individuals of the second brood usually emerge about the middle of March, but the butterflies are very irregular in their appearance at this season. The young larva (Fig. 3b) is much thickened anteriorly, the head being concealed from above by the large thoracic segments. Its colour is pale green, with a pair of long, erect bristles on each segment, a large number of shorter ones being situated on the ventral surface, and behind the head. After the second moult, a brilliant crimson dorsal line is noticeable, but beyond this I have no record, as my larvæ unfortunately died just after completing their third moult. Up to this time they had fed but sparingly on the dock, eating minute holes in the leaves and clinging to them with great firmness. It is much to be regretted that their subsequent history could not be followed, especially as I only succeeded in obtaining the eggs on this one occasion, although I frequently kept females in captivity with this object. Three other species of *Chrysophanus* occur in New Zealand, viz., *C. feredayi*, common round Nelson, and chiefly distinguished by the olive-green undersurface of its hind-wings ; *C. enysii*, which is occasionally met with amongst forest, and may be at once known by its broad black markings and pale yellow colour ; and *C. boldenarum*, a little insect uniting the " Coppers " with the " Blue Butterflies," and found in great abundance in certain river beds and shingly places. The western side of Lake

Wairarapa is one of the best localities I know of for this curious little species.

Family LYCÆNIDÆ.

Lycæna phœbe.

This is the common blue butterfly of New Zealand, which may be observed in great numbers along the roadside on a hot summer's day. Its larva must be very abundant, but has hitherto escaped attention, owing, probably, to its small size. The perfect insect is on the wing from October till May.

Group HETEROCERA.

Family SPHINGIDÆ.

This family is represented in New Zealand by the splendid *Sphinx convolvuli*, an insect I am at present unacquainted with.

Family HEPIALIDÆ.

Porina signata (Plate IX., fig. 2).

Common throughout the summer, when it may be taken in great numbers round lighted windows during any mild evening. The larva is as yet unknown, but is in all probability subterranean in its habits, and feeds on the roots of plants. A large *Hepialus* larva I once discovered under a stone, whilst looking for Coleoptera, was very likely referable to this insect, but as it unfortunately died shortly afterwards it is impossible to speak with any degree of certainty at present. Two closely allied species are *P. umbraculata*, and *P. cervinata*. The former is rather smaller than *P. signata* and of a more uniform brown, with a white stripe in the centre of each fore-wing, surrounded with darker colouring. The latter is one of the smallest of the family, its size at once distinguishing it from any of the rest.   In colour it is pale brownish with numerous black and white markings, varieties occasionally occurring much suffused with the darker colour. It is rather local, but may be found abundantly in the Manawatu district.

Family HEPIALIDÆ.

Hepialus virescens (Plate IX., fig. 1 ♂, 1a ♀, 1c larva, 1b pupa).

This gigantic insect is seen occasionally in the forest during the early summer.   The larva (1c) tunnels the stems of living trees, feeding entirely on wood which it bites off with its strong mandibles.   The plant most usually selected by the caterpillar is *Aristotelia racemosa*, called by the settlers "New Zealand currant," from its large clusters of rich-looking black berries, which appear in autumn.   Other food-plants are numerous, the black maire (*Olea apetala*) and manuka (*Leptospermum*) being among those more frequently chosen.

This larva, for the most part, inhabits the main stem of the tree, its gallery always having an outlet to the air, which is covered with a curtain of dull brown silk, spun exactly level with the surrounding bark, and consequently very inconspicuous.   These burrows usually run down towards the ground, and are mostly two or three inches from the surface of the trunk.   In some instances the larvæ inhabit branches, in which case, if the branch is of small dimensions, the tunnel is made near the centre.   These remarks only refer to galleries constructed by young larvæ, as the tunnel made by the insect prior to becoming a pupa is of a very complicated character and merits a somewhat detailed description.   It consists of a spacious, irregular, but shallow cavity, just under the bark, having a large opening to the air, which is entirely covered with a thin silken covering, almost exactly the same shape and size as the numerous scars which occur at intervals on the trunks of nearly all the trees. Three large tunnels open into this shallow cavity : one in the centre, which runs right into the middle of the stem, and one on each side, which run right and left just under the bark. These are usually very short, but sometimes extend half-way round the tree, and occasionally even join one another on the opposite side. The central tunnel has a slightly upward direction for a short distance inwards, which effectually prevents it from becoming flooded with water ; afterwards it pursues an almost horizontal course until it reaches the centre of the tree when it appears to suddenly terminate. This, however, is not the case, for, if the gallery floor is carefully examined a short distance before its apparent termination, a round trap-door will be found, compactly constructed of very hard, smooth silk, and corresponding so closely with the surrounding portion of the tunnel that it almost escapes detection. When this lid is lifted a long perpendicular shaft is disclosed which runs down the middle of the tree to a depth of 14 or 16 inches, and is about six lines in diameter. At the bottom of this the elongated pupa (Fig. 1b) sleeps quietly and securely in an upright position, the old larval skin forming a soft support for the terminal segment of the pupa to rest on. The upper end of this vertical shaft is lined with silk, which forms a framework on which the trap-door rests when closed. The lid itself is of a larger size than the orifice which it covers, and this makes it extremely difficult, if not impossible, to force it from the outside, whilst it fits down so closely to the aperture as not to be readily lifted. The object of this most ingenious contrivance is, in all probability, to prevent the ingress of insects, large numbers of spiders, slugs, and various Orthoptera being frequently found in both central and lateral tunnels, but they are quite unable to pass the trap-door. The galleries of different individual larvæ are all wonderfully

alike, the only differences observable being in the length of
the perpendicular shaft and the direction of the horizontal
burrow, which is sometimes curved. These variations are
usually caused by the presence of other tunnels in the tree,
which the larva invariably avoids, although how it can
ascertain that it is approaching another tunnel before
actually reaching it, is hard to understand. As development
progresses in the pupa, it becomes darker in colour, especially on the wing-cases, which in some individuals show
the future black markings of the moth, as early as two
months before emergence. Others remain quite white and
soft, the green wings suddenly appearing through their
cases a fortnight or three weeks prior to the bursting forth
of the imago. Previous to this change the pupa works
its way up the vertical tunnel, lifts the trap-door, which
yields to the slightest pressure from within, and wriggles
along the horizontal burrow until it reaches the air, the
last three or four segments only remaining in the tree.
The thoracic shield then ruptures, and the moth crawls out
and expands its wings in the ordinary way, resting on the
trunk of the tree until they are of sufficient strength and
hardness for flight.

The perfect insect, although it must be common, is very
rarely seen. It is best reared from the pupæ, which can
be often successfully cut out of their burrows and kept
amongst damp moss until they emerge. It appears to
be much persecuted by birds, as we often observe its large
green wings lying about on the ground.[1]

The curious "vegetable caterpillar," which is usually
referred to this species, probably belongs to one of the
larger subterranean larvæ of the family.

[1] For a more detailed account of the metamorphosis of this insect
see *The Entomologist*, vol. xviii. p. 30.

THE LEPIDOPTERA. 73

Family BOMBYCIDÆ.

Nyctemera annulata (Plate IX., fig. 3 ♂, 3a larva, 3b pupa).

This abundant species is usually mistaken for a butterfly by the uninitiated owing to its diurnal habits and conspicuous colouring. Its larva feeds on various plants, the most usual being a light green kind of ivy with yellow flowers, but its original food no doubt consisted of the " New Zealand groundsel " (*Senecio bellidioides*), on which it may now be occasionally taken in wild situations. Its general colour is black, with interrupted dorsal and lateral lines, the ventral surface and connecting membrane between the segments being slate-coloured. In younger larvæ there are also several slate-coloured lines extending the whole length of the insect, and thus dividing the black into squares. Round the middle of each segment, at its greatest circumference, a variable number of brilliant blue warts are situated, and out of these dense tufts of long black hair take their rise. There are, however, no warts along the ventral surface. This description applies very well as a rule, but the larva is subject to many slight variations. It remains in this state for nearly three months, or more, according to the season, and is very common, numbers being found on the different plants which constitute its food. The pupa (Fig. 3b) is of a shining black colour, with many longitudinal rows of small yellow blotches on the abdominal segments ; there is also a stripe of the same colour at the tip of the wing-case. It is enclosed in a slight cocoon, formed of a mixture of silk and hair, and is attached near the ground to any firm object. The moth emerges in the course of a month or six weeks. It is very common, being found profusely in the neighbourhood of its food-plants, and appears in the greatest numbers during the early morning hours in the middle of summer.

For an account of a Dipterous insect, parasitic in the present species, I refer to page 59.

Family PSYCHIDÆ.

Œceticus omnivorus (Plate X., fig. 1 ♂, 1a ♀, 1b larva, 1c ♂ pupa).

This insect is very rarely seen abroad, but can be easily reared from the larva, which feeds on manuka and other plants throughout the year. When very young, and in fact immediately after leaving the egg, it constructs a wide spindle-shaped case, principally composed of silk, with a few small fragments of leaves, &c., attached to the outside. It has a large aperture in front, through which the head and anterior portion of the larva are projected, and a much smaller one at the posterior extremity, which allows the pellets of excrement to fall out of the case as they are evacuated. The body of the enclosed caterpillar is of a light straw colour, the head and three first segments being dark brown, with numerous white markings. The abdominal segments are considerably thickened near the middle of the insect, rudimentary prolegs being present on the third, fourth, fifth, and sixth segments of the abdomen. The anal prolegs are very strong, and are furnished with numerous sharp hooklets, which retain the larva very firmly in its case. As it grows it increases the length of its domicile from the anterior, causing it gradually to assume a more tubular form, tapering towards the posterior aperture, which is enlarged from time to time. The outside is covered with numerous fragmentary leaves and twigs of various sizes, placed longitudinally on the case, and frequently near the anterior aperture, the materials, owing to their recent selection, are fresh and green. The interior is lined with soft, smooth silk of a light brown colour, the thickness of the whole fabric being about the same as that of an ordinary kid glove, and so strong that it is impossible to tear it, or indeed to cut it, except with sharp instruments. The size of the case when the caterpillar is mature varies considerably, ranging from 25 to 30 lines or more in length, and about three in diameter, the widest portion being a little behind the anterior aperture (see Fig. 1b).

During the day the larva closes the entrance and spins a loop of very strong silk over a twig, the ends being joined to the upper edges of the case on each side ; in this way it hangs suspended, the caterpillar lying snugly within. I have often known a larva to remain thus for over three weeks without moving, and afterwards resume feeding as before ; this probably occurs while the inmate is engaged in changing its skin. At night the larvæ may be seen busily engaged : they project the head and first four segments of the body beyond the case, and walk about with considerable rapidity, often lowering themselves by means of silken threads ; the only locomotive organs are, of course, their strong thoracic legs, which appear to easily fulfil their double function of moving both larva and case. If disturbed, these insects at once retreat into their cases closing the anterior aperture with a silken cord which is kept in readiness for the purpose, and pulled from the inside by the retreating larva. This operation is most rapidly performed, as the upper edges of the case are flexible, and thus fold closely together, completely obstructing the entrance. When full fed, this caterpillar fastens its case to a branch with a loop of strong silk, which is drawn very tight, preventing the case from swinging when the plant is moved by the wind, and also rendering the insect's habitation more inconspicuous, by causing it to resemble a broken twig. The anterior aperture is completely closed, the loose edges being drawn together and fastened like a bag. The posterior end of the case is

twisted up for some little distance above the extremity,
thus completely closing the opening there situated. It is
lined inside with a layer of very soft silk, spun loosely over
the sides, and partly filling up each end. In the centre of
this the pupa lies with its head towards the lower portion
of the case, the old larval skin being thrust backwards
amongst the loose silk above the chrysalis. In this stage
of existence the extraordinary sexual disparities, which are
so characteristic of the family, manifest themselves, the
male and female pupæ being very widely different in all
respects. The former is figured at 1c, the female pupa
differing from it in the following particulars. It is much
larger and more cylindrical in shape, the abdomen occupying
nearly the whole of the body, and consisting of nine visible
segments, the terminal one being obtusely conical. The
head and thorax are very rudimentary, more resembling
those of the larva than the male, all the appendages being,
however, reduced to hardly visible warts. In colour it is
pitchy black and shining, and its length is about ten lines.
This insect remains in the pupa state during the winter
months, viz., from May till September. When about to
emerge, the male chrysalis works its way down to the lower
end of the case, forces open the old aperture there, and projects the head, thorax, and upper portion of the abdomen,
the pupa being secured from falling by the spines on its
posterior segments, which retain a firm hold in the silk.
Its anterior portion then ruptures, and the moth makes
its escape, clinging to the outside of its old habitation, and
drying its wings. It is probable that the female insect
does not leave her case, communication with the male
being no doubt effected through one of the orifices, and the
eggs afterwards deposited inside. On one occasion I found
a case full of eggs, containing the shrivelled body of the
female and her old pupa shell, which would seem to con-
firm the above opinion. The perfect insects are drawn at

Figs. 1 and 1a. The male (1) is extremely active, dashing about the breeding cage with great rapidity when first emerged, and rapidly beating his wings to tatters ; but the female (1a) closely resembles a large maggot, all the appendages being completely rudimentary, except the two-jointed ovipositor at the end of her body ; she is incapable of any motion, except a slight twirling of the abdomen, which takes place while the eggs are being laid.[1]

Family NOCTUIDÆ.

Leucania nullifera (Plate XIII., fig. 3, 3a larva).

This large, though dull-looking insect, is occasionally taken at light during the summer and autumn months.

The larva feeds on the spear-grass (*Aciphylla squarrosa*), an abundant plant on the coast hills near Wellington. It devours the soft central-growing point, and its presence in a tussock can be at once seen by a quantity of pale-brown "frass," visible at the bases of the leaves. The formidable spear-like points with which this plant is armed must afford the caterpillar considerable protection from enemies. As a rule a single specimen only is found in each clump of the grass, so that the female probably deposits her eggs singly. This larva is full-grown about August, and may be found feeding in the plants during the autumn and winter.

The pupa state is spent, in an earthen chamber, amongst the roots of the spear-grass, and the moth emerges during the summer.

This species occurs at considerable elevations. I have seen it as high as 4,000 feet in the Nelson province, where its food-plant may also be found.

---

[1] For accounts of parasites and hyperparasites of this insect see pages 60 and 37, also *The Entomologist*, vol. xviii. p. 153.

Family NOCTUIDÆ.

Leucania atristriga (Plate X., fig. 2).

Abundant among various blossoms during the latter end of summer, being one of the last of the Noctuæ to disappear in the autumn.

The larva probably feeds on grasses, but I have not yet met with it.

The illustration (Fig. 2) is taken from the male insect, the female differing only in having her abdomen rounded at the tip, a sexual distinction which holds good throughout the family.

Family NOCTUIDÆ.

Erana graminosa (Plate X., fig. 5, 5a larva).

This beautiful insect occurs commonly on the white rata blossoms (*Metrosideros scandens*) round Wellington during March and April, at which time it may be readily taken just after dark with a lantern and killing-bottle. The larva (Fig. 5a) feeds on the mahoe (*Melicytus ramiflorus*) in the spring and autumn. It remains concealed in crevices in the bark during the day, not infrequently selecting the deserted burrows of wood-boring beetles as a secure retreat from its enemies. When full grown it is olive-green, the colour being lighter on the ventral surface and between the segments. A row of ill-defined, feathery, black markings extends down the back and sides and there are also two tolerably conspicuous ochreish spots on every segment except the last. The head, legs, and prolegs are reddish-yellow, and the whole insect is more or less spotted with black. Younger larvæ differ in being of a light yellowish-green, with very pale yellow dorsal and lateral lines. A row of black warts, emitting a few bristles, extend round each of the segments, while the head is pale ochreous with a few black dots.

When full-grown this larva descends to the ground, and

forms a slight cocoon in the earth round the roots of the tree, where it is transformed into a very stout, ruddy-brown-coloured pupa, somewhat paler on the wing-cases. The moth emerges in two or three months' time. Its colouring renders it so inconspicuous amongst moss that I have frequently lifted a handful of the latter out of the breeding cage, and only discovered that the insects had emerged by their falling from the moss on to the table. A very noticeable peculiarity in this species is the presence of a fringe of long hairs in a fold on the anterior margin of the fore-wing. This organ emits a fragrant perfume, and is confined to the male sex (Fig. 5). Only one or two other instances of this kind are at present known among the New Zealand moths.

Family NOCTUIDÆ.

Mamestra mutans (Plate X., fig 7, 7a larva, 7b pupa).

This extremely abundant species occurs almost without intermission during the whole of the year. The sluggish larva (7a) feeds on plantain, and is best obtained by overturning logs and stones, when it may be discovered among the grass and other plants growing round their edge. Its head is pale green, with two broad black stripes, and is clothed with numerous short bristles ; the four succeeding segments are of a ruddy-brown colour, considerably wrinkled, the remainder being light green, suffused with a dull, pinkish hue towards the dorsal surface. The markings consist of a triangular black spot on each side of the second to eighth abdominal segments, and a cloudy lateral line of the same colour ; the legs and prolegs being pale green, and the whole insect more or less marbled with black. This description and the figure on Plate X. exhibit the usual peculiarities of the larva, but in some individuals the markings there indicated are quite obsolete, and the

insect is of an almost uniform pale-green colour. When
mature, this caterpillar sometimes constructs a slight cocoon
amongst moss, on fallen trees, but more often buries itself
in the usual manner, the moth appearing in a few weeks'
time.   Nearly all pupæ collected at random in New
Zealand will be found to give rise to either this species
or the one which immediately follows (*Mamestra composita*).
The perfect insect is most abundant in the spring and early
summer, but may be found fluttering round lamps on any
mild night throughout the year.   The sexes differ considerably : the female is greyish white, with faint brown
markings, while the male is dull reddish-brown, with
the markings considerably darker (Fig. 7).   His antennæ
are also slightly pectinated, those of the female being quite
simple.

Family NOCTUID.E.

Mamestra composita (Plate X., fig. 3, 3a larva).

Very common during the spring and autumn in all open
situations.

Its pretty larva (Fig. 3a) feeds on various grasses, and
threatens in time to do considerable damage to pastures.
The head and dorsal surface of the first segment are dark
shining green, with one or two obscure white markings ;
the rest of the body is ornamented with a number of
parallel brown, white, and orange lines, which render the
larva very inconspicuous when amongst the grass.   Sometimes it occurs in great numbers, nearly every blade of
grass having its caterpillar ; in fact this was almost the case
in the Wairarapa valley in the summer of 1886, when the
larvæ must have produced a marked effect on the paddocks.   When full-grown this caterpillar changes into a
light chestnut-brown pupa, which lies on the surface of the
ground amongst the vegetable refuse.   The perfect insect
appears in about a month's time, and if the evening be mild

may be seen flying with great rapidity at dusk ; it may also be readily captured at light. The figure (3) represents the male insect, the female differing only in her simple antennæ.

Family NOCTUIDÆ.

Mamestra ustistriga (Plate X., fig. 6 ♂).

This handsome insect is rather uncertain in its appearance, but is occasionally taken quite unexpectedly at rest on tree-trunks or palings in the daytime. Specimens may also be captured while feeding on the white rata blossoms early in March, where they occasionally occur among the hosts of other Noctuæ. The larva, which feeds on the honeysuckle, is of a pale brown colour, with two obscure darker lines on each side, the under-surface being light slate-colour. The pupa state is spent in the ground, and many fine specimens may be reared from chrysalids picked up while gardening, &c. The sexes of this insect differ considerably in colour : the male is of a pinkish grey with black markings, while the female is of a uniform pale grey, and considerably smaller.

Family NOCTUIDÆ.

Heliothis armigera (Plate X., fig. 4, 4a larva).

This conspicuous insect occurs in great abundance during certain seasons, but is very irregular in its appearance, it frequently happening that only two or three specimens are noticed in a whole year. It is generally seen flying in the daytime, when it delights to suck honey from the flowers of the Scotch thistle, a plant which much overruns the forest lands when first cleared. The larva (Fig. 4a) is a very handsome caterpillar, of a dark brownish black colour, ornamented with yellow subdorsal and lateral lines and numerous streaks and dots of the same hue. The ventral surface is a rich yellowish brown, and the subventral line

7

white, the spiracles being white with black rings ; a reddish blotch also adorns each of the three thoracic segments.   It feeds voraciously on geraniums, tomatoes, peas, and many other garden plants, where it often commits the most serious ravages.   About the end of April it is full-grown, when it descends to the ground and buries itself two or three inches below the surface.   In this situation it is shortly transformed into a pupa, remaining in that state until the following summer, when the moth appears.   The sexes of this insect differ considerably, the male having the fore-wings of a ruddy-brown colour, sometimes inclining to orange, while in the female they are pale ochreish ; both sexes are, how- ever, subject to considerable variation, and the figure (4) is taken from a rather dark male specimen.

Family NOCTUIDÆ.

Plusia eriosoma (Plate X., fig. 8, 8a larva).

An abundant species round Nelson, where almost any number may be taken hovering over flowers on a still summer's evening.   In Wellington it occurs occasionally. The larva (Fig. 8a) is a pseudo-geometer, having twelve legs, and thus showing a strong affinity with the next family.   In colour it is pale green, darker on the dorsal surface than elsewhere.   A white line runs down each side, and the whole insect is covered with black dots and bristles. The colouring of different individuals varies in intensity, and a fainter white line, above the usual one, exists in some specimens.   It feeds on beans, geraniums, and many other imported plants, and is doing much good in the Nelson gar- dens by the havoc which it is committing among the Scotch thistles—weeds equally injurious to the agriculturalist and the gardener, not only crowding out useful plants, but rapidly exhausting the soil in which they grow.   Formerly this insect must have fed exclusively on the New Zealand nightshade (*Solanum aviculare*), on which plant it may still be occasionally found in the forest, where no imported species are available, but, like many other caterpillars in this country, it is forsaking the native vegetation for the European. When full-grown, this larva spins a slight cocoon of white silk, which is generally placed between two leaves. The pupa is of a shiny black colour, the membrane between the segments being reddish-brown. The moth emerges in about three weeks' time. The figure (8) is taken from a female insect, the male being readily distinguishable by two large tufts of hair situated at the end of his body and often very conspicuous. In some cases the wings of the female are considerably lighter than in the illustration, but otherwise the species does not seem to vary. It is the New Zealand representative of the English "Silver Y Moth" (*P. gamma*), no doubt familiar to many of my readers.

Family GEOMETRIDÆ.

Declana floccosa (Plate XI., fig. 1, 1a larva).

I have started the Geometridæ with *Declana* because it exhibits a great many more points in common with the Noctuidæ than does the genus *Acidalia*, which latter is placed at the head of the Geometridæ by some modern Lepidopterists, chiefly, I believe, on account of neuration, a character which if taken alone cannot but produce the most unnatural divisions. The present insect is one of the commonest of the genus, and may often be observed throughout the whole summer resting on the sheltered sides of trees and fences, occasional stragglers being met with as late as the end of May. Its larva is a pseudo-geometer possessing twelve legs (Fig. 1a), and thus almost exactly resembling the caterpillars of the genus *Catocala*, belonging to the Noctuidæ ; the curious filaments on each side of the insect making this likeness still more complete. It feeds on the " New Zealand currant " (*A. racemosa*), from which individuals can be occasionally beaten during the spring and early summer. They are almost impossible to find by searching in the ordinary way, from a habit they possess of clinging firmly to the twigs, which they exactly imitate in colour. When full-grown this caterpillar constructs a small cocoon just below the ground, where it is transformed into a robust-looking pupa, from which the moth emerges in a month or six weeks' time. The sexes of this species may be readily distinguished, the male (Fig. 1) having the antennæ slightly pectinated, while those of the female are quite simple, and her body much more robust. The moth drawn at Fig. 1b has been reared from larvæ exactly resembling those of the present insect, of which it is consequently now known to be only an extreme variety. It was formerly ranked as a distinct species under the name of *Declana junctilinea.*

Family GEOMETRIDÆ.

Chalastra pelurgata (Plate XI., fig. 2 ♂, 2a ♀, 2b larva).

This delicate species may be taken flying about the forest at night, from October till March, but is most abundant on the white rata blossoms during the latter end of summer.

Its caterpillar feeds sparingly on a delicate fern (*Todea hymenophyllioides*) which grows in dark glades in the forest, where the sun seldom or never shines. In colour it is generally dull brown, with a row of green or pale brown lunate spots on each side; on the ventral surface the colour is darker, except on the thorax, where it is green, the legs being also green. There are in addition numerous fine, wavy lines down the back and sides of the larva, and the dorsal surface of the thoracic segments and ventral prolegs are bright reddish brown (Fig. 2b). These larvæ are, however, very variable ; in many the " lunate " stripes are much longer, having a diagonal direction, and thus extending up the sides of the insect towards its dorsal surface, while others have the ventral surface dark green, and additional markings of more or less importance.

When full-grown it spins a loose cocoon of earth and dead leaves, from which the perfect insect emerges in a month or six weeks' time. The sexes are widely different, both being figured on the Plate (Fig. 2 ♂, 2a ♀). I have noticed that at least four females occur to every male, which is a very unusual arrangement, the males being generally much the commoner among the Lepidoptera.

Family GEOMETRIDÆ.

Ploseria hemipteraria (Plate XI., fig. 3, 3a larva).

A curious moth, occurring in some numbers at various blossoms during the summer evenings, but rather uncertain in its appearance. The larva (Fig. 3a) feeds at night on veronica, where it may be often found with a lantern, devouring the flowers and leaves. In colour it is light green with two yellow lines on each side, the dorsal surface being considerably darker, and almost blue. Specimens are not infrequently met with of a uniform dark brown, and the two conspicuous lateral lines are then reduced to a single obscure ochreous band. These caterpillars are very inconspicuous during the daytime, as they remain quite motionless for hours together, sticking straight out from the stems of their foodplant, which they closely resemble. The pupa is unusually robust, and possesses a sharp spine at its extremity. In colour it is pale olive brown, with a pinkish line on each side of the abdomen, the wing-cases being more or less suffused with pink. It is not enclosed in any cocoon, but may be found amongst the dead leaves round the stems of the veronica. The perfect insect appears in about three weeks' time. It is liable to be passed over for a faded leaf, the general outline and colouring of the wings rendering the insect very inconspicuous, especially amongst foliage. The specimens I have reared all closely resemble Fig. 3, so that this insect does not appear at all prone to vary.

Family GEOMETRIDÆ.

Ploseria alectoraria (Plate XI., fig. 4; Plate XIII., fig. 7 larva).

One of our most variable moths, occurring occasionally amongst foliage during the summer, but most abundant on the white rata blossoms in February and March.

The larva feeds on *Pittosporum eugenioides*, where it may be sometimes found in October and November. It has a most wonderful resemblance to the buds of the plant, and can only be dislodged by vigorous beating. It is easily reared in captivity—in fact the female moths may often be induced to lay their eggs and the insect observed through all its stages.

The eggs are very flat, oval, and light green in colour, becoming brown at one end about five days before hatching.

The young larva is pale green with a dull yellowish head. It has no markings until after the first moult when a reddish dorsal line appears. As age advances the larva becomes darker in colour and is ornamented with a series of diagonal yellow stripes. The spiracles and antennæ are pink and very conspicuous. The legs and prolegs are very small, and the latter are bright red in colour ; a fleshy process which projects from the last segment of the larva is similarly coloured. The whole insect is also speckled with yellow. When full-grown this caterpillar is very robust and measures about ten lines in length. The pupa is enclosed in a light cocoon formed of three or four leaves fastened together with silk. It is greenish brown in colour.

The perfect insect first appears in December. It may be observed during the whole of the autumn and occasionally in the winter. As the larvæ grow very slowly I am inclincd to think that the females hibernate and lay their eggs early in the spring (Fig. 4)₄

Family GEOMETRIDÆ.

Sestra humeraria (Plate XI., fig. 5, 5a larva).

This abundant species occurs in large numbers round Wellington, amongst brushwood, whence it may be often dislodged during the daytime, but is most readily procurable in the evening. The larva (Fig. 5a), feeds on *Pteris incisa,*. a pale green fern, growing in many open spots in the forest to a height of three or four feet. Its general colour is dull brownish yellow, slightly darker on the back, and ornamented with a number of wavy yellow lines on each side. The ventral surface and legs are green and the head is dark brown ; the whole insect being covered with numerous black dots and bristles. When disturbed these larvæ immediately drop to the ground, and coiling themselves up like small snakes, become very inconspicuous.

The pupa is buried in the earth about two inches below the surface, the insect remaining in this state during the winter months. The moths generally emerge about October. So far as my experience goes they are not subject to any notable variations. The specimen drawn at Fig. 6 is regarded as a variety of this species by Mr. Meyrick, but I myself believe it to be quite distinct, as among over a dozen *humeraria* larvæ reared in captivity, none of the imagines had the slightest resemblance to Fig. 6, although the caterpillars were all taken within a few yards of the place where such moths occurred.

Family GEOMETRIDÆ.

Selidosema dejectaria (Plate XI., fig. 8 ♂, 8a ♀, 8b larva).

An abundant and conspicuous species, occurring throughout the summer, often noticed at rest on fences and trees during the day and always taken in great numbers on various blossoms in the evening.

The caterpillar is extremely variable, the colouring of different individuals being apparently much influenced by their surroundings ; those specimens, for instance, taken from the pale green foliage of the mahoe (*M. ramiflorus*) resemble in colour the twigs of that plant, while others captured feeding on the white rata (*Metrosideros scandens*) are dark reddish brown.   Fig. 8b is drawn from a larva found on the fuchsia, which, when in its favourite position, viz., sticking straight out from the side of a branch, is so much like one of the sprouting twigs that it absolutely defies detection.   When full-grown this insect buries itself about two inches in the earth, where it shortly becomes a dark chestnut-brown pupa, lighter between the segments. The time required for the development of the perfect insect depends upon the season, larvæ which undergo their trans-formations in the spring developing much more rapidly than those that feed up in the autumn.[1]

This insect is extremely variable, having been formerly divided into several distinct species; the two most usual forms are those shown at Figs. 8 and 8a, but every intermediate variety exists.   The sexes are distinguished by the usual differences in the antennæ.   My experience leads me to believe that the light varieties occur more frequently in the female than in the male sex, and also that the dark larvæ give rise to dark moths, and *vice versâ*, although a great many more specimens will have to be reared before these can be regarded as established facts.

---

[1] On one occasion I enclosed a full-grown caterpillar of this in-sect in a pot of earth with a recently formed Noctua pupa, whose internal portions it immediately devoured, employing the empty shell of the unfortunate chrysalis as a cocoon. It is impossible to say whether this horrible proceeding often occurs in a state of nature.

Family GEOMETRIDÆ.

Selidosema panagrata (Plate XI., fig. 7 ♂, 7a ♀, 7b larva).

One of our commonest moths, occurring in great numbers in the forest throughout the whole summer.

The larvæ (Fig. 7b) are extremely variable, the most usual colouring being that of the individual figured, but when very young they are all of a uniform green with a con-spicuous white dorsal line ; as age advances the cater-pillars become dark olive brown of varying degrees of intensity in different specimens, some retaining a consider-able amount of their original green colouring, especially those feeding on the kawakawa (*Piper excelsum*), whose hue consequently harmonizes with that of the plant. These larvæ often select a forked twig to rest in, where they lie curled round with the head and tail close together. They are very voracious, and are the primary cause of the riddled appearance which the leaves of the kawakawa almost invariably present. Other food-plants are the "currant" (*A. racemosa*), and the *Myrtus bullata ;* those taken from the latter have a strong pinkish tint, and are consequently very inconspicuous amongst the young shoots where they generally feed. The burrows of *Hepialus virescens* are frequently utilized by the larvæ which feed on the "currant," as convenient retreats during the winter, a large number being often found in a single hole. When full-grown they descend to the ground and construct, on the under-side of fallen leaves, loose cocoons of silk and earth from which the perfect insects emerge in about a month's time. The autumnal larvæ, however, either hibernate or remain in the pupa state throughout the winter. This moth is even more variable than the last species (*S. dejectaria*), which it occasionally somewhat resembles. The sexes are very different, the colouring of the male consisting of various shades of warm brown (Fig. 7), while in the female the prevailing hue is slaty brown or even grey (Fig. 7a). Many specimens are much suffused with ochre and reddish-brown, while the stigma near the centre of the forewing, although sometimes almost obsolete, is often very conspicuous and black, white, or even yellow in colour. It would be of great interest to learn, by rearing a large number of these insects, whether the many varieties existing in the larval and perfect states could be traced to differences in food-plant, or some other external circumstance.

Family GEOMETRIDÆ.

Selidosema productata (Plate XII., fig. 1 ♂, 1a ♀, 1b larva).

Abundant in the forest, where it may be dislodged from ferns and undergrowth during the day or captured flying about in the evening. Its larva is rather attenuated, and possesses a large hump on the second abdominal segment. In colour it is dark reddish brown, mottled with creamy white and pale green, and is sparsely supplied with a few isolated hairs (Fig. 1b). It feeds on the white rata (*Metrosideros scandens*), and when in its usual position—*i.e.*, sticking straight out from a branch—absolutely defies detection. Specimens, however, may be readily procured with a lantern at night, when they may be found walking about and eating. The pupa state is spent in the earth, about two inches below the surface, the moth appearing in three or four weeks' time, this period, however, being extended in the case of autumnal larvæ, to as many months. It is extremely variable, scarcely two individuals being found exactly alike. The colouring, as in the caterpillar, is chiefly protective, consisting of a delicate tracery of browns and greys, which render the insect quite invisible when resting on the trunk of a tree, with its pale yellowish hind-wings concealed, a position it invariably assumes during the daytime (Fig. 1 male, 1a female). The curious and interesting "*Tatosomas*," with their enormously elongated bodies, are closely allied to the present insect; one of them (*Tatosoma agrionata*) being found in similar situations, although in much more limited numbers; as, however, I know nothing of their transformations, I am forced reluctantly to pass them by.

Family GEOMETRIDÆ.

Hydriomena deltoidata (Plate XIII., fig. 1, 1a larva).

One of our commonest moths, appearing in great numbers during January and February, in all open situations. It is especially abundant on the fern-hills.

The larva (Fig. 1a) feeds on the plantain. It is very sluggish, and lives all through the winter, becoming full-grown in September, when it changes into a pupa, among the roots of its food-plant. In colour it is a uniform dark brown.

The moth is extremely variable, but the figure may be taken as representing a fairly typical specimen. It is a pretty insect, and may be often seen resting on fences with its fore-wings folded backwards and forming together a triangle, whence its name of *deltoidata*. Any unusual-looking specimens of this species should always be netted, in order to form a thoroughly representative series, as many of the varieties are very interesting. A rather uncommon and remarkable-looking form occasionally occurs, in which the dark central band of the fore-wings is completely divided near the middle.

Family GEOMETRIDÆ.

Asthena schistaria (Plate XII., fig. 2, 2a larva).

This delicate little insect may be often taken at rest on fences and tree-trunks during the day, and is a con-

spicuous moth when flying in the evening, owing to its
light colour. The larva (Fig. 2a), which feeds on the
manuka (*Leptospermum ericoides*), is very ornamental. Its
general colour is light green, with black dorsal and lateral
stripes, and a series of diagonal markings bordered with
crimson ; the legs and prolegs are also crimson, and the
segments are divided by brilliant yellow rings, a white line
extending down each side of the larva. It is difficult
to find, as it remains closely concealed amongst the dense
manuka foliage, from which it can only be dislodged
by vigorous and continued beating. The caterpillars
allow themselves to fall a short distance, hanging suspended by a silken thread, which they rapidly ascend
when the danger is passed. The pupa is rather attenuated,
dark-brown, and much pointed at its posterior extremity.
It is found buried about an inch in the earth, and the
moth appears in a month's time. This insect varies
much in intensity of markings. The males are generally
considerably darker than the females, but are more cer-
tainly distinguished by their attenuated bodies.

The pearly white *Asthena pulchraria* occurs in October
and April ; it is a most beautiful insect, and may be found
amongst the foliage of the kawakawa (*P. excelsum*), on
which its larva will probably be found to feed.

Family PYRALIDÆ.

Scoparia hemiplaca (Plate XII., fig. 4).

This pretty little moth was reared from a larva
found feeding amongst moss during the winter of 1885,
but unfortunately I neglected to make a drawing until
it was too late. Doubtless many of the other Pyrales we
meet with in the New Zealand forest have similar habits,
their larvæ probably feeding on different kinds of mosses.
These can always be examined during the winter months,

when the entomologist is usually in want of work, and thus much information may be obtained regarding this interesting but little-known family.

Family PYRALIDÆ.

Scoparia sabulosella (Plate XIII., fig. 4, 4a larva).

This is that extremely abundant, though dull-coloured little insect, that rises in such multitudes from every field before one's footsteps during the early summer.

Its larva (Fig. 4a) feeds on various mosses, forming numerous silken galleries amongst the roots in which it resides. These caterpillars are very active, and consequently rather difficult to obtain, as they move either backwards or forwards in their galleries with equal rapidity.

They feed during the whole of the autumn and winter, changing into pupæ about September, from which the moths emerge in a month or six weeks' time.

The habits of the numerous other species belonging to this genus and the closely allied genus *Xeroscopa* (Meyr) probably do not materially differ from those of the species here described.

Family PYRALID.E.

Crambus flexuosellus (Plate XII., fig. 5).

An extremely abundant insect, occurring in swarms over meadows during the summer, where it may be captured in the daytime or taken by hundreds at the attracting lamp in the evening. Its larva is at present unknown, but probably feeds on the roots of grasses.

Closely allied is *Crambus tahulalis*, found in similar situations, but appearing rather later in the season, the earliest specimens being met with about January, while *C. flexuosellus* is on the wing throughout the summer.

Family PYRALIDÆ.

Siculodes subfasciata (Plate XII., fig. 3, 3a larva, 3b pupa).

This curious insect may be occasionally taken flying round patches of *Muhlenbeckia adpressa*, which grows freely amongst brushwood in many parts of the country.

Its larva (Fig. 3a), is very stout and sluggish, resembling the caterpillar of an ordinary Pyrale in general appearance. It feeds in the stems of the creeper, causing large swellings therein, which readily betray its presence, and should therefore be cut off and kept until the moth emerges, as specimens obtained in this way are far superior to any captured in the open. The pupa is dark brown, and shining; it lies in the centre of one of the swellings, the larva having previously prepared a safe outlet for the moth in the form of a small burrow leading to the air, its extreme end remaining closed by a thin pellicle of the original bark, which effectually prevents the inmate's resting-place being discovered from the exterior (see Fig. 3b, the small circle marked * represents the outlet).

The perfect insect appears about December, flying rapidly in the hottest sunshine. It varies greatly, both in size and colour, some of the small males being very much suffused with dark brown, while the females usually resemble the figure (3), and are often more than twice the size of their mates. This insect is generally placed in a family called the *Siculidæ*, but I think without sufficient reason, and have therefore located it among the Pyralidæ, with which it has unquestionably a great affinity.

Family TORTRICIDÆ.

Isonomeutis amauropa (Plate XIII., fig. 2, 2a larva).

This odd little moth may be occasionally seen basking in openings in the forest, and usually flies away with lightning speed when an attempt is made to capture it.

The larva lives under the scaly bark of the matai-tree, feeding on the soft, juicy inner bark and sap. In colour it is light yellowish white, darker on the back, some specimens becoming quite pink on the dorsal surface. When full-grown it encloses itself in a tough silken cocoon, covered on the outside with fragments of wood, from which the moth emerges in about a fortnight's time.

The sexes differ considerably in appearance, the male having much broader wings, and darker in colour than those in the female from which the illustration (Fig. 2) is taken.

This insect is probably single-brooded, as the larva may be found feeding in the trees during the whole of the winter.

Family TORTRICIDÆ.

Cacoecia excessana (Plate XIII., fig. 5, 5a larva).

This is the commonest species of *Tortricidæ* in New Zealand, and may be found almost without interruption during the whole of the year.

The larva (Fig. 5a) feeds on a great variety of plants, the common manuka being probably the most usual food for the species when in a state of nature It now, however, eats numerous European plants, including honeysuckle and occasionally the fruit of the apple, but further evidence is required on the latter subject before we can really consider it as actually injurious in that direction.

In colour this caterpillar is light green with a yellow line on each side, but varies considerably ; it feeds between several rolled-up leaves, in which it is afterwards converted into a pupa whence the moth emerges in about three weeks' time.

The perfect insect is also excessively variable and is often more or less suffused with yellow. It is most abundant in

the middle of summer, and may be taken at light, or in
the daytime at rest on fences and trees.

Family TORTRICIDÆ.

Ctenopseustis obliquana (Plate XII., fig. 6).

This little moth is occasionally noticed at rest on garden
fences during the autumn. Its larva inhabits the interior
of the peach, feeding on the kernel, which appears to exactly
meet its requirements, the caterpillar being full-grown
as soon as it has completely devoured the nut. Before
assuming the pupa state this insect provides a ready means
of escape for the future moth by drilling a small hole
through the hard shell and pulp of the peach to the air ;
it also spins a slight cocoon inside the stone, the pupa
resting in the place formerly occupied by the kernel, in
which position it is often discovered. The only noticeable
mischief produced by this insect is delay in the ripening
of the fruit. In fact all the infected specimens which I
have seen were quite hard and green, whilst other fruit
from the same tree had reached complete perfection.

Family TINEIDÆ.

Endrosis fenestrella (Plate XII., fig. 7, 7a larva, 7b pupa).

This common species may be observed in almost any
house in New Zealand, and is often mistaken for the
dreaded " clothes moth " (*Tinea tapezella*), which it somewhat resembles in general appearance. Its larva (Fig. 7a) is
very destructive, feeding on dried peas, amongst which it
creates great havoc, drilling numerous holes through them
and spinning a large number together, in the centre of
which the caterpillar undergoes its change into a pupa
(Fig. 7b), from which the moth emerges in about a fortnight's time. This insect should be destroyed whenever
seen, as there is no doubt that much loss will be caused by
its ravages in the future. It also infests bee-hives.

Family TINEIDÆ.

Œcophora scholæa (Plate XIII., fig. 6, 6a larva).

This dull-coloured insect. is extremely abundant during the early summer.

The larva feeds on the roots of various plants, forming numerous white silken galleries in the earth where it resides. In colour it is dark chocolate-brown with a yellowish head and white markings. It is very large, considering the size of the future moth, full-grown specimens often measuring as much as 10½ lines in length. About the end of September these caterpillars are transformed into pupæ, and the moths emerge in a month or six weeks' time.

The perfect insect may be often disturbed amongst brush-wood. It is very sluggish on the wing and usually drops to the ground, where it is very inconspicuous. It also has a habit of running into any crevice immediately on the approach of an enemy. This peculiarity is shared by the other members of the genus *Œcophora*, of which there are large numbers in New Zealand.

Family TINEIDÆ.

Semiocosma platyptera (Plate XII., fig. 8, 8a larva, 8b pupa).

This is one of the largest of the *Tineidæ* found in New Zealand, measuring fully fifteen lines across the expanded wings. Its larva (Fig. 8a) is abundant under the bark of dead henau trees (*Eleocarpus dentatus*), feeding on the soft inner surface, but leaving the hard wood untouched. In colour it is pale yellow, the head and prothorax are dark brown and corneous, and the remaining segments are provided with two horny warts, from which numerous hairs arise ; its legs are all very small, and the caterpillar is considerably attenuated posteriorly ; it is very active, wriggling about with great violence when disturbed.

The pupa (Fig. 8b) is enclosed in a compact cocoon, constructed of minute fragments of wood, firmly woven
together with silk, and attached to the inner surface of the
bark, where it may be soon found by careful searching,
and the finest specimens may thus be easily reared in
captivity.

The perfect insect appears about November, and may be
often observed at rest on the trunks of trees; its pale hind-
wings are completely concealed by the dark upper pair,
which render its discovery very difficult. The sexes may
be at once distinguished by their size, the males being much
smaller than the female (Fig. 8) and usually lighter in
colour.

CHAPTER VII.

The Neuroptera.

The Order Neuroptera, as here considered, is a very limited one, consisting only of the seven small families, which comprise the Lace-wings, Ant-lions, Caddis-flies, and a few others. It forms a most convenient passage from the insects undergoing a complete metamorphosis with a quiescent pupa, to those which are active during the whole of their life, as the larvæ are widely different from the adults, but the pupæ, although incapable of walking or eating, approximate very closely in structure to the perfect insects. I regret that my observations have been at present restricted to three families only, *i.e.*, the *Hemerobiidæ*, *Sialidæ*, and *Phryganidæ*, which will consequently have to represent the entire series. I understand, however, from Mr. A. S. Atkinson, that a species of *Myrmeleontidæ* (Ant-lion) is not uncommon round Nelson, and doubtless future investigation will reveal insects belonging to the other families.

Family PHRYGANIDÆ.

Oxyethira albiceps (?) (McLach.) (Plate XIV., fig. 3, 3a larva, 3b pupa).

This insect occurs in the neighbourhood of ponds and streams during the summer. Its larva may be found commonly in the green, slimy weed floating in large masses on all stagnant waters. Being very small it is rather difficult to detect, and is best procured by washing a small quantity of the weed in a saucer of water, when the little insects will be at once seen walking about at the bottom. On examination with the microscope the case will first arrest attention, being of a most unique structure. Its shape is best described as closely resembling that of a minute pocket-flask, very much flattened at the lower end and almost transparent. Its surface is slightly corrugated, and the neck of the flask constructed of a much denser material than the body. It is open at both ends, the posterior end being perforated by a long shallow slit, which extends for nearly the whole width of the case, thus admitting a free circulation of water round the larva, which is also able to turn round and project its head and anterior segments through the lower aperture, thus occupying the reverse position to that shown in the illustration (Fig. 3a). It is, however, prevented from actually leaving the case by its abdomen, which is too large to be withdrawn from either end. The head and thorax of the larva are very horny in comparison with those portions permanently retained in the case, the legs being constructed to fold up into the smallest possible compass, a cavity existing in each joint for the reception of the preceding one—a structure which is almost universal among the caddis-worms. The two organs, situated on the posterior segments, are doubtless respiratory in their function, a large air-tube taking its rise from each and ramifying through the body in all directions. When alarmed these insects retreat into their cases with lightning rapidity, remaining concealed until the danger is passed. Their food probably consists of the green weed, although they are perhaps carnivorous, feeding on the rotifers and other animalculæ, which swarm in the water where they are found.

With regard to the method employed by the young larva in constructing, and subsequently enlarging, its case, I can give no positive information, although it is undoubtedly made of a viscous fluid, secreted by the insect, which hardens when exposed to the water ; this secretion is no doubt analogous to the silk of caterpillars, which always exists in the form of a gummy fluid before being spun.

When about to change, the insect fixes its case down by four ligaments, two at each end, the extremities of these being firmly fastened to a stone ; it then closes the small aperture, and constructs a curious arch-shaped partition, of dense material, a short distance from the broad end (Fig. 3b). In about a week's time the larva is transformed into a pupa, having the limbs, &c., free from the body but incapable of motion. The fixing down of the case prior to the change may be easily performed from each of the apertures, which are no doubt left open till the last for this purpose. Before the final transformation the pupa breaks through the partition at the broad end of the case and rises to the surface, the imago (Fig. 3) ascending a blade of grass to dry and expand its wings. The little exuvia of the pupa may be often noticed floating on the water, and the empty cases are very conspicuous on the sides of a glass aquarium, where the insects generally fix them down when in captivity.

Family HEMEROBIIDÆ.

Stenosmylus incisus (Plate XIV., fig. 2).

This lovely insect is figured as an example of this family, being found occasionally in the New Zealand forest, but is rather scarce as a rule. I regret that nothing is at present known of its transformations.

Family SIALIDÆ.

Chauliodes diversus (Plate XIV., fig. 1, 1a larva, 1b pupa).

During still warm weather, from December till March, this large insect is frequently observed flying lazily over water at dusk, when it may be readily captured with the ordinary net. Its larva is aquatic, living under stones in running streams, where it devours large quantities of Ephemeræ and other insect larvæ, which are always abundant in those situations. It is very ferocious and will bite violently when disturbed, being furnished with a pair of powerful mandibles. The curious filaments on each side are gills, and it will be noticed that they are situated exactly where the spiracles of the perfect insect afterwards appear (see Fig. 1a).

This larva probably lives over a year, its growth proceeding very slowly, but mature specimens are not infrequently met with quite as large as the illustration. When full-grown it leaves the water and forms an oval cell in the mud, usually under a large stone ; its gills then gradually shrivel up, and in ten days or a fortnight it is transformed into the curious pupa, shown at Fig. 1b, from which the perfect insect proceeds in about six weeks' time. The sexes of this species may be readily distinguished by their size, the male being considerably smaller than the female (Fig. 1), and possessing longer antennæ.

CHAPTER VIII.

𝕮𝖍𝖊 𝕺𝖗𝖙𝖍𝖔𝖕𝖙𝖊𝖗𝖆.

THIS Order, although including a comparatively small number of species, comprises some of the largest and most conspicuous insects inhabiting New Zealand, many of them reminding one of the denizens of the tropics in their gigantic size and striking appearance. They may be conveniently divided into the three following groups :— The *Aquatic group*, or those whose larvæ inhabit the water, including the Dragonflies, Mayflies, and Perlidæ ; the *Terrestrial group*, including all the typical Orthoptera, Termites, and Mallophaga ; and the *Euplexoptera*, including the Earwigs. We start our observations with the Aquatic group, as these exhibit the greatest affinity with the Neuroptera.

AQUATIC Group.

Family LIBELLULIDÆ.[1]

Uropetala carovei (Plate XV., fig. 1 ♂, 1a larva.)

This magnificent insect occurs in all swampy situations during January and February, when it may be seen dashing about with amazing rapidity intent on catching

---

[1] The *Libellulidæ, Ephemeridæ, Perlidæ, Psocidæ,* and *Termitidæ* are usually included in the *Neuropteria.*

the various flies which constitute its food. Its curious larva is represented at Fig. 1a, the drawing having been taken from a singularly perfect exuvia, which I had the good fortune to discover, clinging to the stem of a fuchsia-tree in a swamp, the rent through which the perfect insect escaped having almost closed up. In this state it no doubt feeds on various aquatic animals, which it procures with a pre-hensile instrument similar in structure to the "mask" of British dragonfly larvæ, but much larger.

The female of this species may be at once recognized by the absence of the two peculiar leaf-like appendages at the anal extremity, from which the insect takes its name. Her abdomen is also much stouter. My experience leads me to believe either that she is very retired in her habits or else that there are at least six males to one female.

Closely allied, and much commoner than the above insect, is *Cordulia Smithii*, found almost everywhere, its rapid and continuous flight frequently taking it many miles away from any water. The specimen figured is a male (Plate XV., fig. 2), the female possessing a pair of slender sickle-shaped hooks, attached to the end of her body. She may occasionally be seen depositing her eggs in stagnant streams, the abdomen being violently beaten against the surface of the water during the operation. I have not yet met with the larva, which probably lives concealed in the mud. One specimen, taken near Lake Wairarapa, is remarkable in possessing a cloudy brown patch near the tip of each wing, but it is no doubt only a variety of the ordinary insect.

Family LIBELLULIDÆ.

Lestes colensonis (Plate XV., fig. 3, 3a larva).

Extremely abundant in all damp situations from September till May, being one of the last insects to disappear in the autumn. The larva is found under stones, &c., in every stream, feeding on various aquatic insects and crustaceans. When very young the wing-cases are scarcely discernible, but gradually become more distinct at each moult, until the larva assumes the form shown in the illustration (Fig. 3a), which is taken from a specimen about a week before the emergence of the perfect insect. In all these insects it would be much more convenient to regard the metamorphosis as consisting of only two stages, viz., larva and imago, as there is really no condition analogous to the quiescent pupa of other orders. The female is rather stouter than the male, which is the sex figured, and her abdomen is of a dull bronze colour, instead of metallic blue. The only other dragonfly found in my neighbourhood (Wellington) is the pretty little *Telebasis zealandica* (Fig. 4), which occurs in similar situations to the last, but is not quite so common. The male is of a brilliant red colour, the female being bronzy green, but she may be readily distinguished from the same sex in *Lestes colensonis* by her smaller size. The larva of this species is rather more attenuated than that of the previous insect, and is of course considerably smaller.

Family EPHEMERIDÆ.

Ephemera, n.s., near *Coloburus*[1] (Plate XVI., fig. 4, 4a larva).

The well-known mayflies are very extensively represented in New Zealand, hovering in swarms over running water during the summer evenings.

The larva of the present species (Fig. 4a) occurs abundantly under stones in rapid streams. It may be immediately distinguished from its numerous congeners by its large head and conspicuous black eyes. It is carnivorous,

---

[1] One mutilated ♀ specimen of this insect was sent to Mr. McLachlan, but was too imperfect to describe from.

feeding on various small insects, chiefly those belonging to the present family, but in lack of these it will even devour individuals of its own species. It is consequently a most difficult insect to rear, and it was a long time before I succeeded in obtaining a single imago in captivity. When mature the insect leaves the water, and an apparently perfect imago escapes through a rent in the thorax in the usual way. In a few hours, however, a second moult occurs, the wings gaining additional size and beauty, and the anal setæ becoming very much more elongated than before (Fig. 4). This second change, which has so perplexed some entomologists, is merely an *apparent* departure from the general rule, a careful examination of the exuviæ of the dragonflies, and pupa shells of many other insects, revealing a delicate membrane within, which invests the imago, and is cast off at the same time as the harder external envelope. In the case of the mayflies, the retention of this internal membrane some two or three hours longer than usual, will fully explain its apparently unique metamorphosis.

Family PERLIDÆ.

Stenoperla prasina (Plate XVI., fig. 3, 3a larva).

This is the green gauzy-winged insect which we see flying feebly over running water, during the twilight, throughout the summer.

Its larva (Fig. 3a) is aquatic, hiding itself under stones, and devouring the unfortunate *Ephemeræ* found in similar situations. Towards the end of its career the rudimentary wings become very conspicuous, at which time it is a most interesting object. The curious appendages on each side of the abdomen are gills, which the larva is constantly vibrating, in order to obtain a fresh supply of aërated water. When mature, it ascends the stem of some aquatic plant, the skin becomes dry and brittle, and finally bursting, allows the perfect insect to escape, and in a few hours its wings are sufficiently hardened for flight. Several other species occur in New Zealand, one of the commonest being *Perla cyrene*, a black insect much resembling *S. prasina*, but considerably smaller ; its larva may be occasionally found, and is at once known by its dark colour.

TERRESTRIAL Group.

Family PSOCIDÆ.

Psocus zealandicus, n.s. (Plate XVI., fig. 2, 2a larva).

During the hottest days in summer every one must have noticed numbers of minute active insects assembled on garden fences in groups, ranging from ten to fifty, immediately dispersing when disturbed. These are individuals of *Psocus zealandicus* (Fig. 2), a curious little species, closely allied to the renowned " Book Tick " (*Atropos pulsatorium*), whose ravages in museums and libraries need no description. Its larva (2a) may be found in the same situations as the imago, and often assembles in similar groups. Its food probably consists of rotten wood and other decaying vegetable matter, and in its later stages it is provided with wing-cases, thus differing from the Book Tick (*A. pulsatorium*), which remains apterous during the whole of its life.

Family TERMITIDÆ.

Stolotermes ruficeps (Plate XVI., fig. 1 ♂, 1a ♀, 1b "soldier," 1c " worker ").

The termites, or white ants, which occur in such great numbers in the tropics, are represented in New Zealand by several small species, the commonest in this neighbourhood being *Stolotermes ruficeps*.

This species inhabits rotten logs, excavating extensive burrows, resembling in a very humble manner the wonderfully elaborate nests constructed by the African and other species, about which so much has been written, and so much remains to be discovered. The present insect appears in the perfect state during January and February. It is seldom noticed flying about, but may be readily obtained by opening the nests, where a large number are frequently seen huddled together in the main galleries. At this time the community consists of three classes of individuals, viz., males, females, and workers, which last are in all probability nothing more than the larvæ. After pairing they shed their wings and return to the nest, the female becoming very much distended with eggs. About March she commences to lay. This is continued for several months, and during this time the female is queen of the nest. She resides in a capacious chamber, from which numerous galleries diverge in all directions, some extending as far as eighteen or twenty inches, but the most populous portion of the nest is contained within a radius of six inches from the queen's apartment. The "soldiers" (Fig. 1b) now appear in considerable numbers. They are chiefly stationed in the royal chamber, and furiously attack any intruders; but the workers which stream in and out, carrying the eggs from the queen, they treat with the greatest gentleness. I have never seen soldiers in a nest containing winged insects, nor indeed later in the spring than October, when they seem to have all disappeared. With regard to the nature of these individuals I am unable to supply any positive information, but it appears probable that they are abortive males, in the same way that the neuters of the bees and ants are abortive females. As none of these insects have yet been reared, many points of great interest remain to be discovered in connection with their economy, and a rigid investigation of a number of nests kept in captivity, is the only mode by which we can hope to become fully acquainted with the habits of this interesting family.

Family BLATTIDÆ.

Periplaneta fortipes (Plate XVII., fig. 5).

Few people who cut up old wood remain unacquainted with this species for very long, its insufferable odour immediately betraying its presence independently of anything else. It is very common under the bark of rimu, henau, and other large trees, where specimens may be found in all stages of growth ; the mature individuals only differing from the young in the matter of size and the possession of rudimentary wing-cases. I have never found the females of this species carrying their eggs, but have, on several occasions, discovered the closely allied, but smaller, *Periplaneta undulivitta* thus engaged under stones on the hills round Nelson. This is a much more agreeable insect to study than *P. fortipes*, not possessing the disgusting odour so characteristic of the latter species.

The only winged *Blattidæ* found round Wellington are *Blatta conjuncta*, and *Periplaneta orientalis*. The former (Fig. 6), may be occasionally noticed under the scaly bark of rimu and matai trees, but a sharp eye and hand are needed to effect a capture, the insect running with marvellous rapidity. The latter species I have not yet noticed, but as it is the ordinary " cockroach " of Europe its habits have already been amply described.

Family MANTIDÆ.

Tenodera intermedia (Plate XVII., fig. 2).

A local species confined, I believe, to the South Island, and occurring in some numbers round Nelson, where my specimens were obtained. It seldom flies, but crawls stealthily about the trunks of trees, in the hottest sunshine, capturing and destroying great quantities of insects, its green colouring and leaf-like form rendering it very inconspicuous

to its victims.   The purple spots on the tibiæ of this insect
are very noticeable, and resemble small drums in structure,
hence they are regarded by Mr. A. H. Swinton ("Insect
Variety," page 239), as the organs of hearing.   These curious
drums may be also found in insects belonging to nearly all
the remaining families of the Orthoptera, but, as we find no
auditory organs occupying a similiar situation in any other
groups of insects, I think that Mr. Swinton's explanation
of their function must be regarded at present as a
somewhat doubtful one.[1]

Family PHASMIDÆ.

Acanthoderus horridus (Plate XIX.).

The curious Stick Insects are familiar to most people
from their remarkable similarity to the twigs of trees.

The present species is one of the largest, the mature insect
frequently attaining a length of five inches.   It is best
taken at night, when it may be readily discovered, feeding
on the leaves of shrubs, and suddenly becoming perfectly
motionless when the lantern is turned upon it.   The favourite plant for this (and indeed most of the species) is the
white rata, upon which they are often seen in large numbers
when the entomologist is collecting Lepidoptera in autumn.
One of the commonest species found in this way is *Bacillus*
(*hookeri ?*) chiefly remarkable for its great sexual disparities,
the male resembling a very slender stick about twenty-eight
lines long, while the female is nearly half as long again
(thirty-eight lines), and much more stoutly built.   A
more systematic investigation of this family is needed
before we can pretend to correctly determine the various
species, as there is little doubt that in other cases the
sexes will be found quite as divergent.   In addition to this

---

[1] For account of the earlier stages of this, or a closely allied insect,
see "Transactions of New Zealand Institute," vol. xvi. p. 114.

the insects are most variable in colour, and their completely
apterous character rendering the distinction between larva
and imago a matter of considerable difficulty, it is
very probable that some of the smaller species may be
only immature specimens of the larger ones.

Stick insects are easily kept in captivity, and will not be
found devoid of interest. They are great eaters, and
grow with considerable rapidity, frequently casting their
skin, a task of no easy accomplishment, which I once had
the pleasure of watching in the case of a specimen of *Acanthoderus prasinus* which I had under observation for
several months.

The insect first suspends itself by its hind pair of legs,
keeping the others in the same position as when walking,
the head is bent in, and the antennæ are placed along the
breast, the long abdomen hanging over backwards.
The skin then splits along the back of the thorax, and
the head and thorax are gradually pushed out. The
front and middle legs are immediately afterwards extracted, the long femora and tibiæ easily passing the
sharp angles in the exuvia, owing to their complete
flexibility. When these are finally clear, the insect reaches
forwards with its fore-legs and draws the abdomen and
hind-legs out of the old skin, which remains attached to
the branch until dislodged by some accident.

During the spring months great quantities of little stick
insects may be noticed on the parasitic ferns covering the
tree stems in the forest ; they are curious little animals,
their antics when simulating inanimate twigs being often
most amusing, and if the reader wishes to investigate a
comparatively untouched branch of entomology he cannot
do better than keep a number of these until mature, when
he will doubtless contribute much to our scanty knowledge
of this curious family.

Family ACHETIDÆ.

Acheta fuliginosa (Plate XVIII., fig. 1).

This destructive insect is not indigenous to New Zealand, having been introduced from Australia into the Nelson district many years ago. Strange to say it has never been seen in Wellington, where specimens must be constantly landed amongst produce, &c., but appear to be unable to effect a settlement, owing, probably, to some peculiarity of the . climate which renders the place unsuitable for them. The larvæ may be first observed about December, when they are often seen hopping about the vegetation. They are extremely obnoxious, devouring everything, and frequently entering houses, where they consume provisions, clothes, and even boots. During the summer of 1875 the farmers round Nelson were fairly eaten out by this insect, the cattle absolutely starving for the want of food, but since that time the pest seems to have gradually diminished, although it is still very injurious to many garden plants.

The illustration (Fig. 1) is taken from a female, the male wanting the long ovipositor. These insects appear in the imago state about March, and continue in great abundance until the end of summer, the cold weather which generally sets in about the beginning of May rapidly destroying them.

Family GRYLLIDÆ.

Deinacrida megacephala (Plate XVIII., fig. 2 ♂; XVII., fig. 8 ♀).

This conspicuous species is especially interesting, as it may be regarded as the type of a very peculiar assemblage of apterous crickets, pre-eminently characteristic of New Zealand. It is very abundant round Wellington, and may be occasionally taken under logs, &c., but is best procured

from the hollow stems of various trees, where it is found
inhabiting the deserted galleries of wood-boring species—
frequently enlarging them to suit its own requirements.

The plant most usually selected by these insects is the
mahoe (*Melicytus ramiflorus*), whose stems may be often
seen pierced with large holes. Out of these the insects
emerge at night to feed on the leaves. To extract a
number of specimens, without injury, requires considerable
care, and is best performed with a small axe, which should
be first used to cut in about three-quarters through the
trunk, just below one of the holes. Another notch is then
cut about a foot lower down, and the intermediate wood
split off in long pieces, until the tunnel is laid bare. On
approaching an insect the first thing seen are two red
threads, which are the antennæ, laid back as shown at
Fig. 8. A deep notch is then cut into the trunk, some
nine or ten inches below this point, and the piece bodily
wrenched off. If the individual thus treated is a male he
will cling firmly to the log, elevating his hind-legs in
the air and biting viciously at anything within reach, but
the females, in the majority of cases, endeavour to escape
and hide themselves under the leaves, &c., on the ground.
Both sexes when irritated emit a peculiar grating sound,
which may be often heard at night in the forest, and is
produced by the friction of the femur against a small file
situated on each side of the second abdominal segment.
They can also leap a short distance, but not so far as
many of the smaller species (*Libanasa macropathus*, &c.).
They are evidently strictly arboreal in their habits, as
they exhibit great skill in walking along branches, and
will climb up a thin stick with wonderful rapidity.

When in their burrows the posterior legs are extended
behind the insect and push, while the anterior and intermediate ones are thrust forwards, the claws being firmly
inserted, so as to enable the insect to pull itself along.

Travelling along the burrow in this manner, they frequently evade all efforts to extract them, until they are stopped by arriving at the end of the gallery.

The sexes of this species are readily distinguishable, the male (Plate XVIII., fig. 2) possessing an immense head furnished with a pair of enormously powerful mandibles. The female (Plate XVII., fig. 8) is a more attractive insect, her gracefully curved ovipositor and smaller head having a much more pleasing appearance than the terribly menacing jaws of her mate. Both sexes are able to give severe bites, but it is extremely doubtful whether they would prove anything worse than slight mechanical injuries, as the insect is not likely to be poisonous. I am, however, unable to tpeak from experience.

Family GRYLLIDÆ.

Xiphidium maoricum (Plate XVII., fig. 1).

This pretty insect may be found in great abundance round Nelson during the autumn, but is rarer in the Wellington Province. Its presence may be at once detected by the curious chirping heard in various directions shortly before sunset and lasting till eight or nine o'clock in the evening. This sound is produced with the wing-cases, which the male insects may be seen vigorously rubbing together. The females are quite mute, and they may be also distinguished by possessing a short curved ovipositor at the end of the body. The peculiarly leaf-like shape of the insect and its bright green colour render its discovery amongst the herbage a most difficult matter, even when its whereabouts is indicated by its cry—in fact, were it not for their music, there is little doubt that very few of these insects would ever be captured, as they are practically invisible, and are an instance of protective resemblance carried to great perfection.

When disturbed these crickets fly about twenty yards

and again settle in a bush or amongst herbage, carefully avoiding alighting on the ground where they would be readily visible. Their flight is somewhat feeble for such large insects. Great care must be taken, when capturing specimens for preservation, not to hold them by their powerful hind-legs, as they will not infrequently cast one off while endeavouring to escape.

I have not yet noticed the larva of this species, but should imagine it would closely resemble a wingless imago.

Family LOCUSTIDÆ.

Caloptenus marginalis (Plate XVII., fig. 4).

This is the little grasshopper which rises before our footsteps in swarms on a hot summer's day ; it is one of the last insects to leave us in the autumn, being frequently found in warm situations on fine days in the middle of winter. Owing to its great abundance this species must inflict considerable damage on the grass, as it has taken up its quarters like the English grasshopper in the cultivated fields, where an unlimited supply of food is always at hand. Formerly, no doubt, it was much less common round Wellington than at present, owing to the few open spots then existing, none of these grasshoppers being found in the forest.

The perfect insect may be recognized by the rudimentary wings which are present on the thorax, thus causing it to closely resemble the larval form of many of the winged species, and for which it might readily be mistaken were its true character unknown.

Family LOCUSTIDÆ.

Œdipoda cinerascens (Plate XVII., fig. 3).

This large and conspicuous insect occurs abundantly in all open situations near Nelson, but is very rare in the

Wellington district, becoming, however, again common further north.

When disturbed it leaps into the air, spreads its wings, and flies away with great rapidity for thirty or forty yards, when it alights, and allows its pursuer to get within a few yards of his prize before again making off. This habit renders the capture of a good series of this insect a most arduous matter. The sexes may be readily distinguished by their size, the female being nearly twice as large as her mate.

This species is very variable in colour, some individuals being dark green whilst others are of a uniform drab.

The food of this insect consists of various domestic grasses, but I do not think it is at present sufficiently abundant to exercise any harmful influence on agriculture. By some entomologists, however, it is regarded as only a variety of the renowned migratory locust (*Locusta migratoria*), and as such its advent in large numbers might be viewed with serious apprehension.

It is also strange that although I have often seen large numbers of this species in the perfect state I have never observed the larva. I can only conjecture that the insect breeds in very secluded localities and then migrates in search of fresh food supplies.

Group EUPLEXOPTERA.

Family FORFICULIDÆ.

Forficesila littorea (Plate XVII., fig. 7).

Abundant on the sea beach throughout the year, where it may be readily captured under stones and seaweed. It is a very bold insect, and when disturbed will grasp a blade of grass, or other object, very firmly with its powerful abdominal forceps, and allow itself to be lifted off the ground and carried away rather than relinquish its hold.

The food of this species probably consists of seaweed, although it is possibly carnivorous, and feeds on the small insects and crustaceans, which are numerous on the beach. Being permanently apterous, mature individuals can only be recognized by their large size, and the perfect development of their anal forceps. It is evidently erroneous to regard these as organs exclusively employed in opening and shutting the wings, as we see that in the present insect, which does not require them for that purpose, they are larger than in many of the flying earwigs. They are probably chiefly used to *intimidate* intruders.

This species is strictly marine in its habits and is seldom found more than a few yards above high-water-mark. The females may be often observed hatching their eggs. For this purpose they excavate an oval chamber underneath a log or large stone, and after carefully smoothing it within, deposit the eggs at the bottom. These eggs are most faithfully guarded by the mother, which boldly attacks all intruders, and will suffer herself to be killed rather than leave the spot. She also remains with the young ones for a considerable time after they are hatched, as we sometimes observe the females accompanied by a number of larvæ of quite a large size.

CHAPTER IX.

The Hemiptera.

THE present Order of insects, although of very limited extent, contains several important species, of which the noisy Cicadas, destructive Aphides, and numerous Bugs, and Lice, can be cited as familiar examples. The Hemiptera may be conveniently divided into the two following groups :—

The *Homoptera*, comprising all the species in which the anterior wings are entirely membranous, and—

The *Heteroptera*, including those having the basal portion of the anterior wings thickened, and quite opaque.

These peculiarities have induced some entomologists, who regard the structure of the wings of the greatest importance in classifying, to arrange the insects included in the Homoptera and Heteroptera, into two distinct Orders ; but their uniform character in all other respects renders this, I think, hardly desirable.

Group HOMOPTERA.

Family CICADIDÆ.

Cicada cingulata [1] (Plate XX., fig. 1, 1a pupa).

This beautiful insect may be found in great numbers

[1] This genus is frequently called Melampsalta.

amongst brushwood during the hot sunny days so common from January till March. Its larva inhabits the earth earlier in the summer, and its curious pupa can often be observed crawling up the stems of trees in order to allow the perfect insect to emerge. After this has taken place the exuviæ still remain firmly attached to the tree, and are very conspicuous objects ; but if it is desired to remove them great care must be taken not to break off the legs, which are always very brittle.

The perfect insects are at once betrayed by their loud singing, which, in certain localities, becomes quite deafening. This noise is entirely confined to the males, and proceeds from two large drum-like organs, situated on the under surface of the abdomen near its base, which, in conjunction with the curious ovipositor existing in the females constitute good sexual distinctions throughout the family. The structure of these two organs having been admirably described by several European authors renders it quite unnecessary for me to do so here.

Closely allied to the present insect is *Cicada muta*, the female of which is depicted on Plate XX., fig. 2. The male is often of a reddish-brown colour, but the insect is an extremely variable one. It is found in similar situations to *C. cingulata*, but appears rather earlier in the year.

Family CICADIDÆ.

Cicada iolanthe, n.s. (Plate XX., fig. 3, 3a larva, 3b pupa).

This is the first species of Cicada to appear in the spring, and is found during November and December. Its larva (Fig. 3a) is a curious little animal, the two hindlegs being very long. I am at present unable to state with certainty what constitutes its food, but am extremely doubtful whether it consists of the juices imbibed from the roots of plants, as is generally supposed. The anterior legs, although probably chiefly constructed for digging, appear to be also suited for raptorial purposes, which leads me to believe that the insect may be carnivorous in its habits. The pupa (Fig. 3b) does not materially differ from that of the last, except in size, and its empty exuvia is also frequently found attached to the stems of trees.

The perfect insect may be at once discovered by the peculiarly shrill note emitted by the male.

Family APHIDÆ.

This family is extensively represented in New Zealand, but as I have not yet been able to obtain any information respecting their specific identity I am compelled to pass them by for the present, hoping that future investigation will reveal much that is interesting in their habits, and also help both gardener and agriculturist to protect himself from their ravages.

Family COCCIDIDÆ.

Coelostoma zealandicum (Plate XX., fig. 4 ♂).

This species is figured as a representative of this very curious family chiefly on account of its great similarity to a Dipterous insect, the rudimentary condition of its posterior wings being most perplexing to the beginner. Its habits have been amply described by Mr. Maskell, in his work on the Coccididæ of New Zealand, to which I consequently refer.

Group HETEROPTERA.

Family NOTONECTIDÆ.

Corixa zealandica, n.s. (Plate XX., fig. 5).

Abundant throughout the summer in all slow-running streams. The larva closely resembles the imago except that it has no wings. Its food probably consists of the juices of other insects. The present insect invariably swims with

its back exposed, thus differing considerably from the
English Water-boatman (*Notonecta glauca*), whose keel-like
back is kept beneath the water, while the two long hind-
legs are rapidly moved backwards and forwards like oars.

Family SCUTELLERIDÆ.

Cermatulus nasalis (Plate XX., fig. 6, 6a larva).

This insect may be beaten out of various trees during
the summer, and is usually taken in some abundance in
February amongst white rata blossoms, on which it may
be often observed sucking the honey from the blossoms
with its long rostrum. Its larva, which is represented at
Fig. 6a, is found in similar situations.

This concludes the series of insects I have selected as
representative of the several orders in New Zealand. The
brief sketch of entomology thus given is of necessity extremely fragmentary, and many important groups and
families are entirely unrepresented. Should, however, this
little book induce some of its readers to investigate insects
for themselves, I shall feel that my efforts have been amply
rewarded.

THE END.

GENERAL INDEX.

EXPLANATION OF PLATES.

Note.—In all the Plates and references thereto the sign ♂ indicates that the specimen figured belongs to the male sex, ♀ to the female sex, and ☿ to the neuter sex.

In the case of enlarged figures the insect's natural size is indicated by a line.

PLATE I.

COLEOPTERA.

Fig. 1.—Cicindela tuberculata.
   „ 1a.—Larva.
   „ 2.—Chætosoma scaritides.
   „ 3.—Pterostichus opulentus.
   „ 3a.—Larva.
   „ 4.—Colymbetes rufimanus.
   „ 4a.—Larva.
   „ 5.—Staphylinus oculatus.
   „ 6.—Dryocora howittii.
   „ 6a.—Larva.
   „ 7.—Dorcus punctulatus.
   „ 8.—Stethaspis suturalis.
   „ 8a.—Larva.

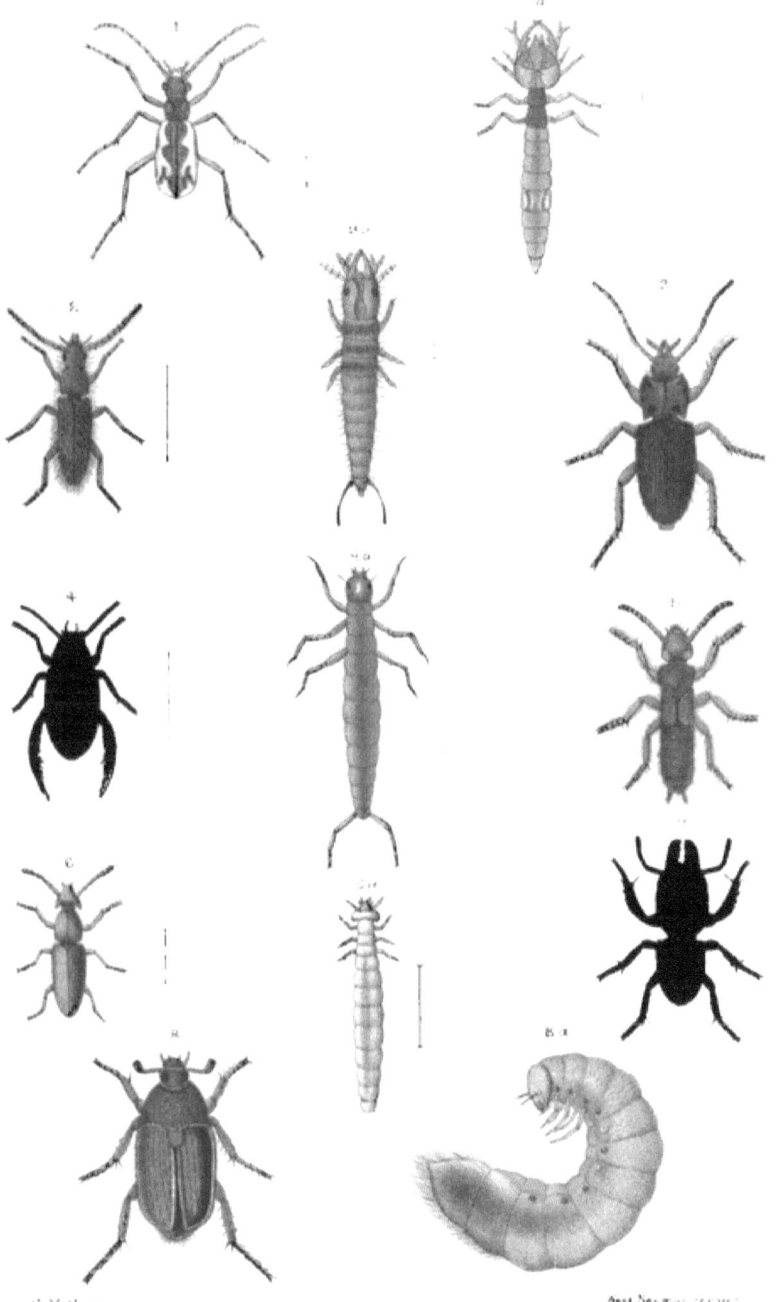

PLATE II.

Plate II.

West, Newman chromo

PLATE III.

HYMENOPTERA.

Fig. 1.—Dasycolletes hirtipes. (?)

  „ 2.—Pompilus fugax.

  „ 3.—Formica zealandica ♂.

  „ 3a.— „   „   ♀.

  „ 3b.— „   „   ☿.

  „ 3c.—Cocoon.

  „ 4.—Ponera castanea ♂.

  „ 4a.— „   „   ☿.

  „ 4b.—Larva.

  „ 5.—Atta antarctica ♂.

  „ 5a.— „   „   ♀

  „ 5b.—Larva.

  „ 6.—Ichneumon sollicitorius.

  „ 7.— „   deceptus.

  „ 8.—Scolobates varipes.

  „ 9.—Pteromalus (?), n.s.

  „ 10.—Dasycolletes purpureus.

Plate III.

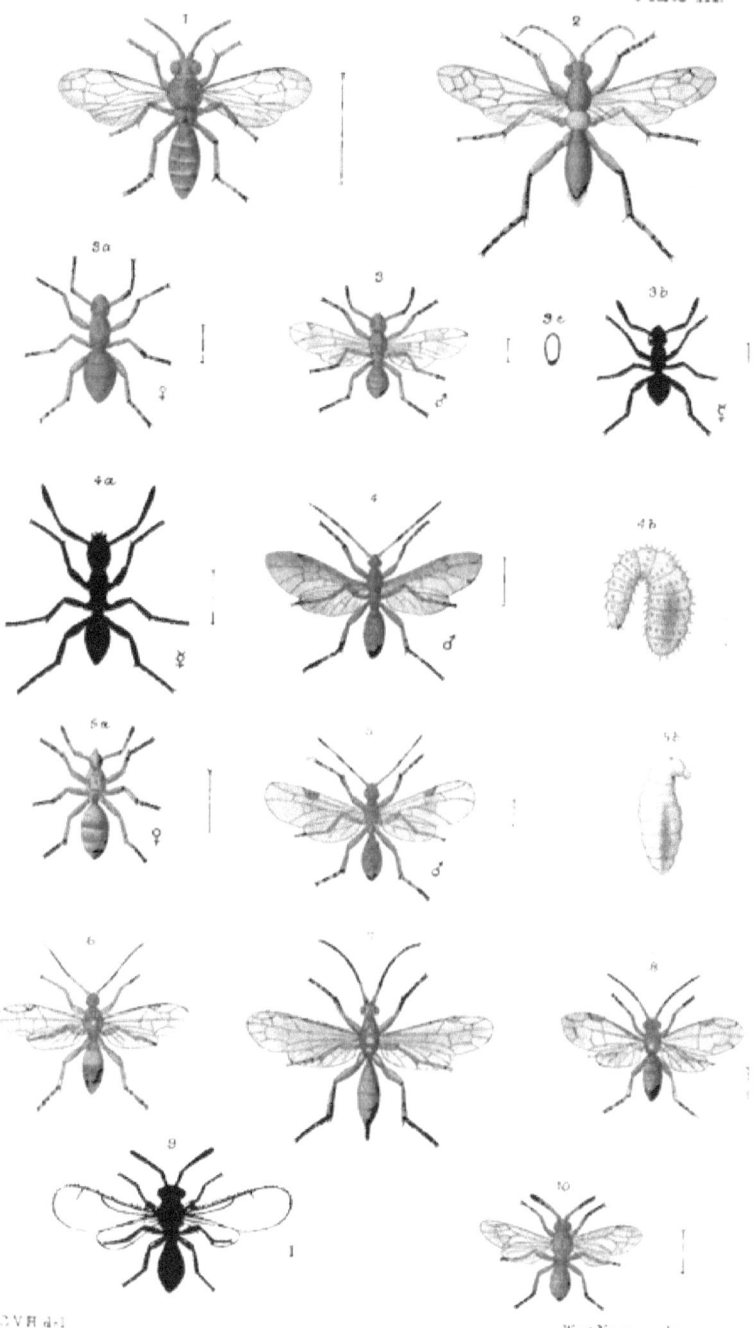

PLATE IV.

DIPTERA.

Fig. 1.—Culex iracundus ♀.
,, 1a.—Larva.
,, 1b.—Pupa.
,, 2.—Chironomus zealandicus, n.s.
,, 2a.—Larva.
,, 2b.—Pupa.
,, 3.—Corethra antarctica, n.s.
,, 3a.—Larva.
,, 3b.—Pupa.
,, 4.—Ceratopogon antipodum, n.s.
,, 4a.—Larva.
,, 4b.—Pupa.
,, 5.—Mycetophila antarctica, n.s.
,, 5a.—Larva.
,, 5b.—Pupa.
,, 6.—Psychoda conspicillata.

Plate IV

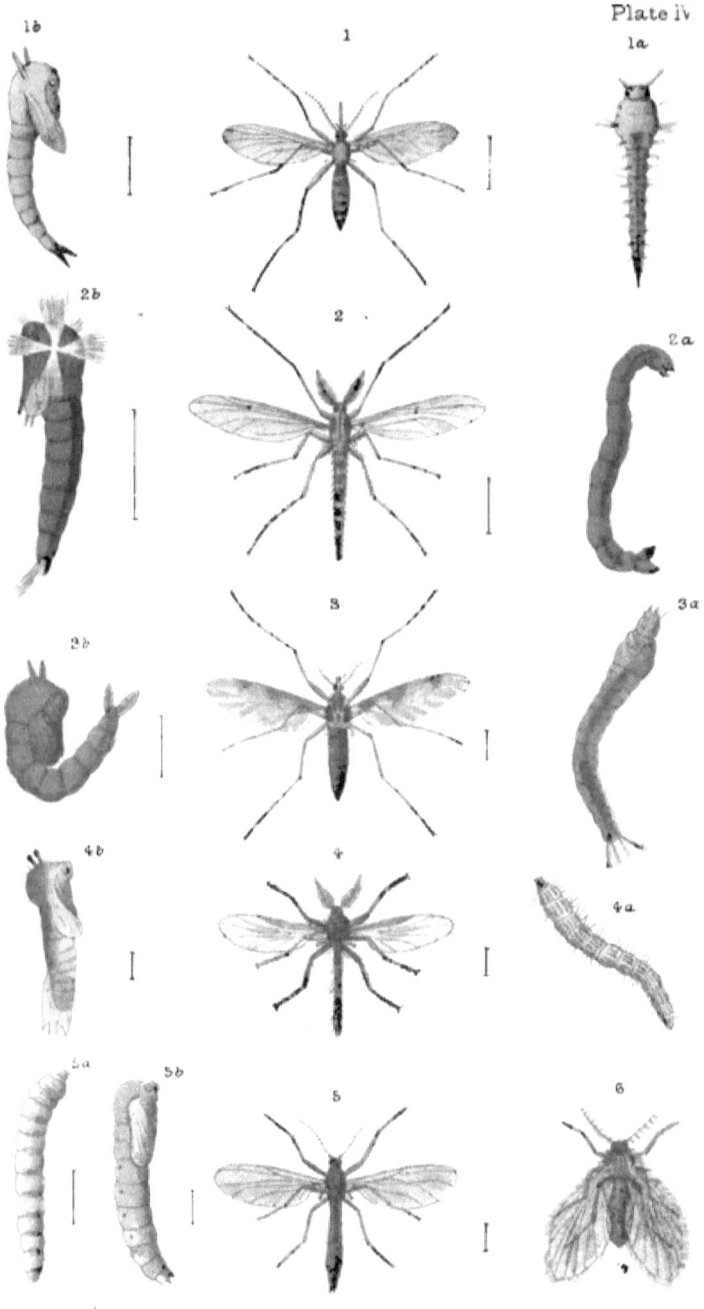

1b 1 1a
2b 2 2a
3b 3 3a
4b 4 4a
5a 5b 5 6

G V H del.

West Newman chromo.

PLATE V.

DIPTERA (*continued*).

Fig. 1.—Tipula holochlora.
„ 1a.—Larva.
„ 1b.—Pupa.
„ 2.—Tipula fumipennis, n.s.
„ 2a.—Larva.
„ 2b.—Pupa.
„ 3.—Cloniophora subfasciata.
„ 3a.—Larva.
„ 4.—Rhyphus neozealandicus.
„ 4a.—Larva.
„ 4b.—Pupa.
„ 5.—Bibio nigrostigma ♂.
„ 5a.—Larva.
„ 5b.—Pupa.

Plate V

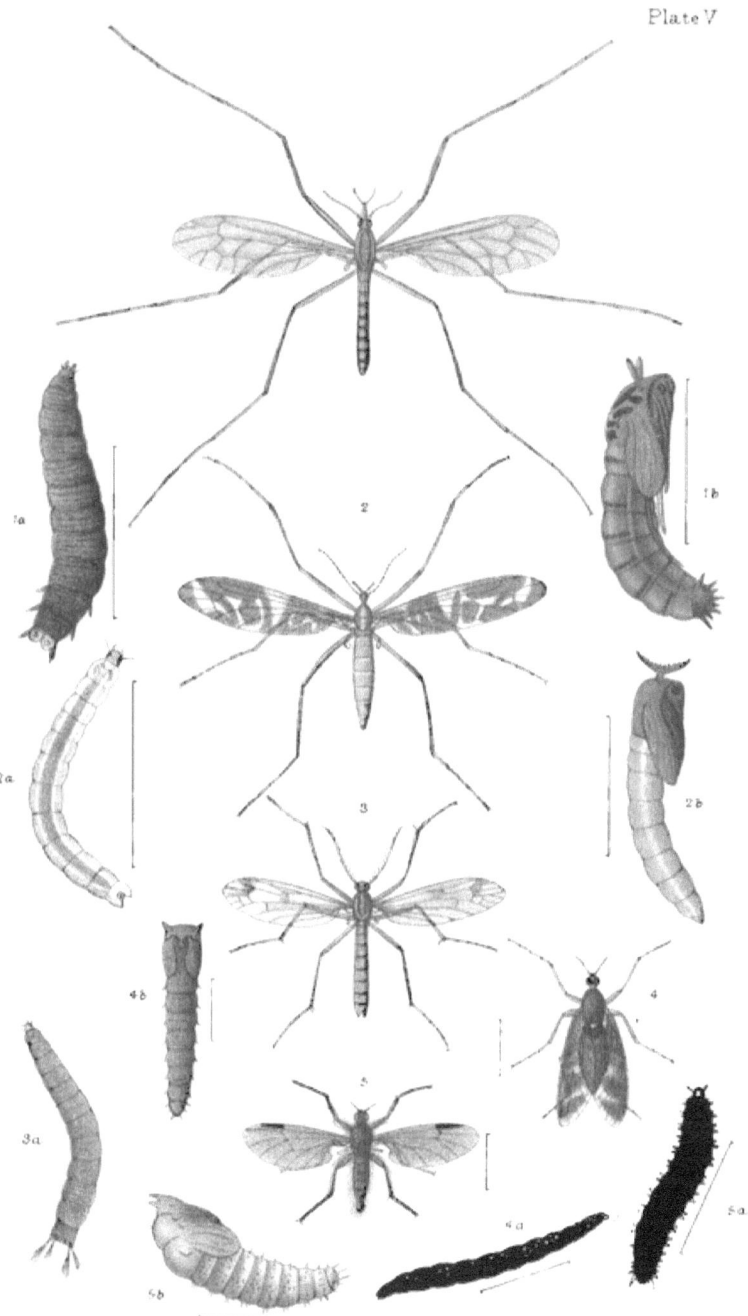

PLATE VI.

Diptera (*continued*).

Fig. 1.—Simulia australiensis.
„   1a.—Larva.
„   1b.—Pupa.
„   2.—Comptosia bicolor.
„   3.—Comptosia virida, n.s.
„   3b.—Pupa.
„   4.—Sarapogon viduus.
„   4a.—Larva.
„   4b.—Pupa.
„   5.—Exaireta spiniger.
„   6.—Tabanus impar.

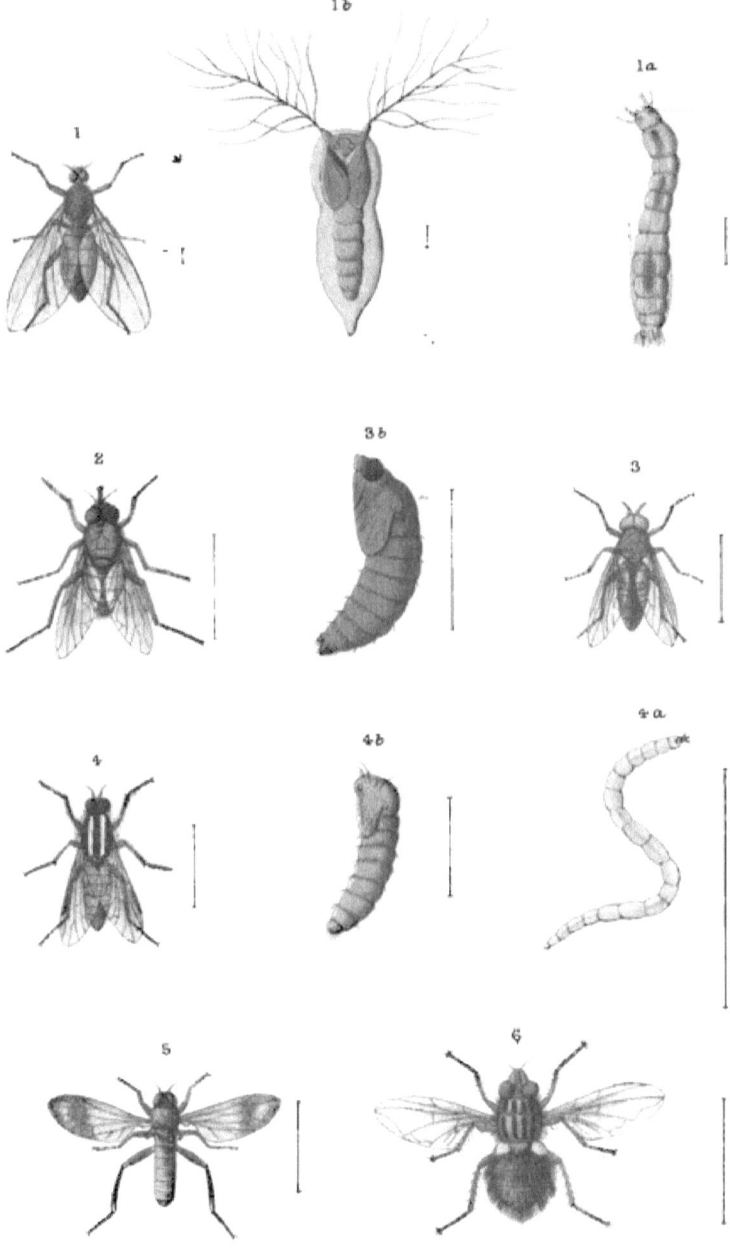

PLATE VII.

Fig. 1.—Helophilus trilineatus.
" 1a.—Larva.
" 1b.—Pupa.
" 2.—Eristalis cingulatus.
" 3.—Syrphus ortas.
" 3a.—Larva.
" 3b.—Pupa.
" 4.—Acrocera longirostris, n.s.
" 5.—Miltogramma mestor ?
" 6.—Nemorea nyctemerianus, n.s.
" 7.—Eurigaster marginatus.
" 9.—Calliphora quadrimaculata.
" 10.—Sarcophaga læmica.
" 12.—Œstrus perplexus, n.s.
" 13.—Cœlopa littoralis.
" 14.—Cylindria sigma.
" 15.—Phora omnivora, n.s.
" 15a.—Pupa.

Plate VII

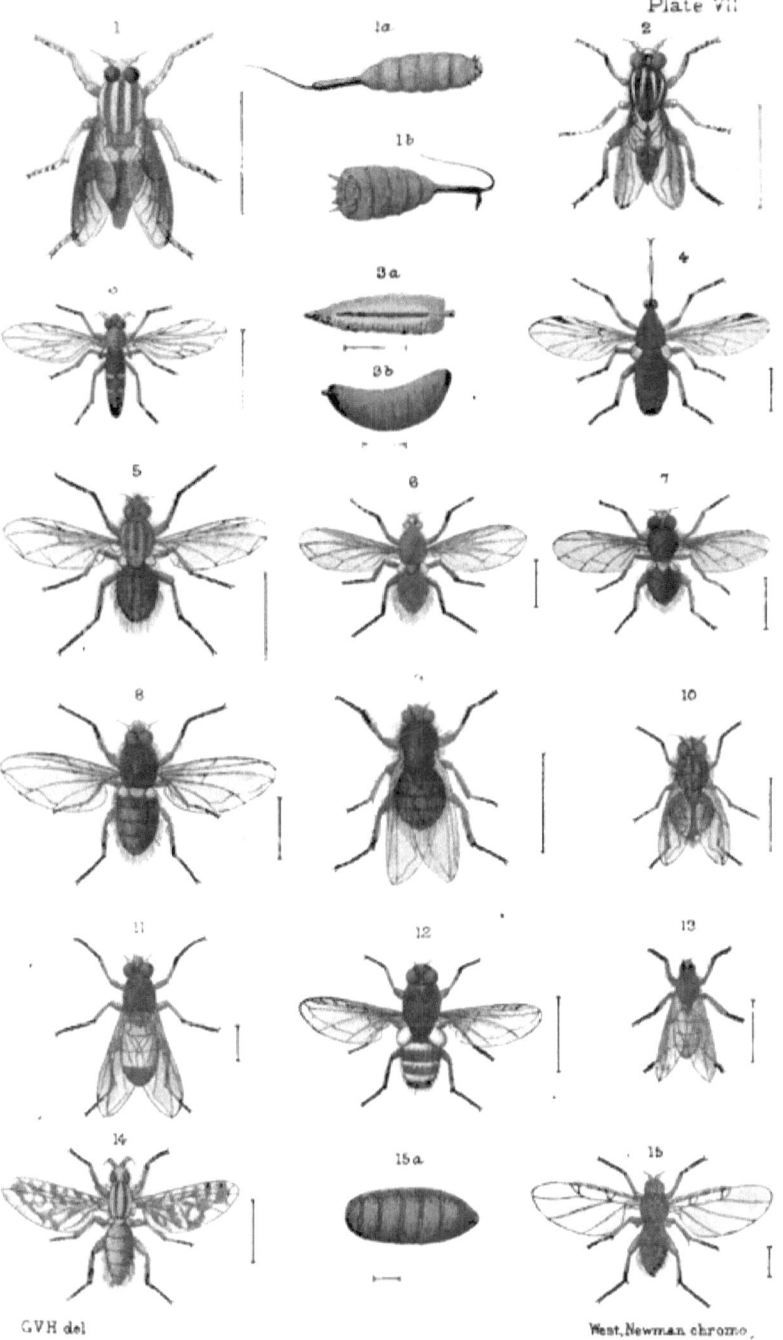

PLATE VIII.

LEPIDOPTERA.

Plate VIII

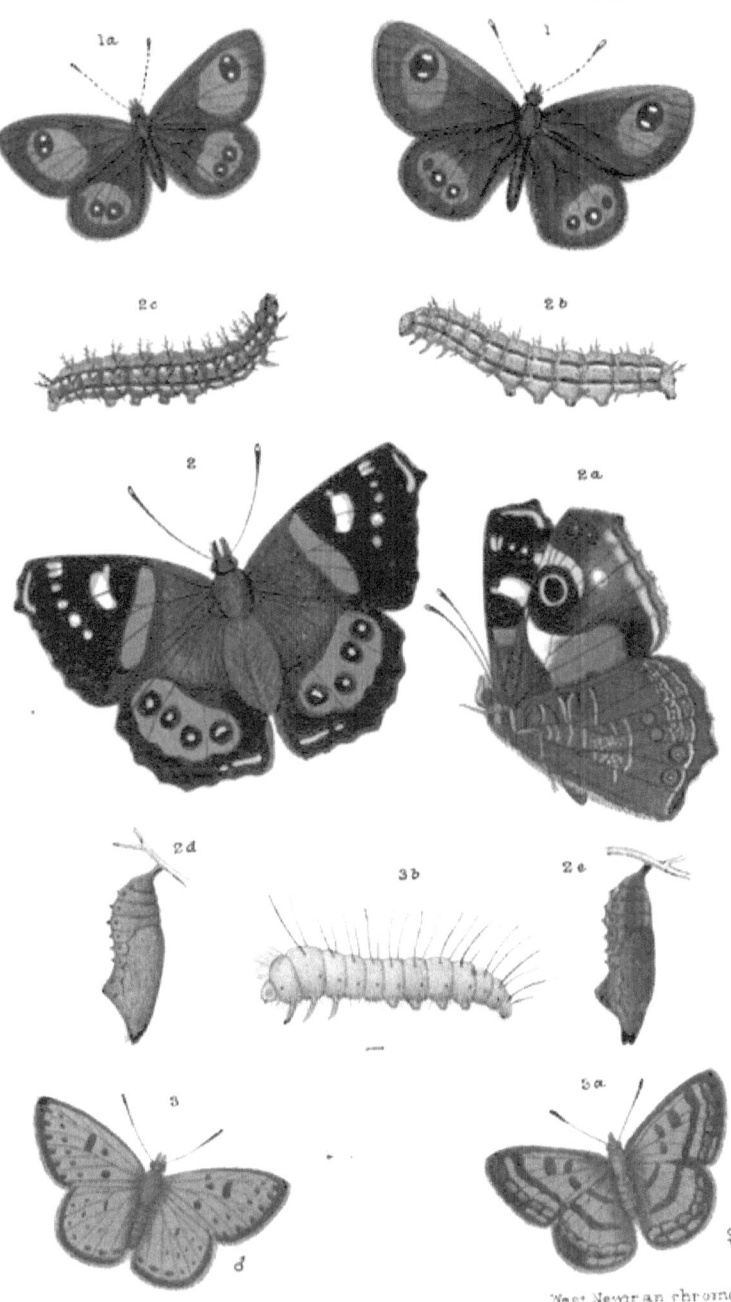

C.V.H. del

PLATE IX.

Plate IX

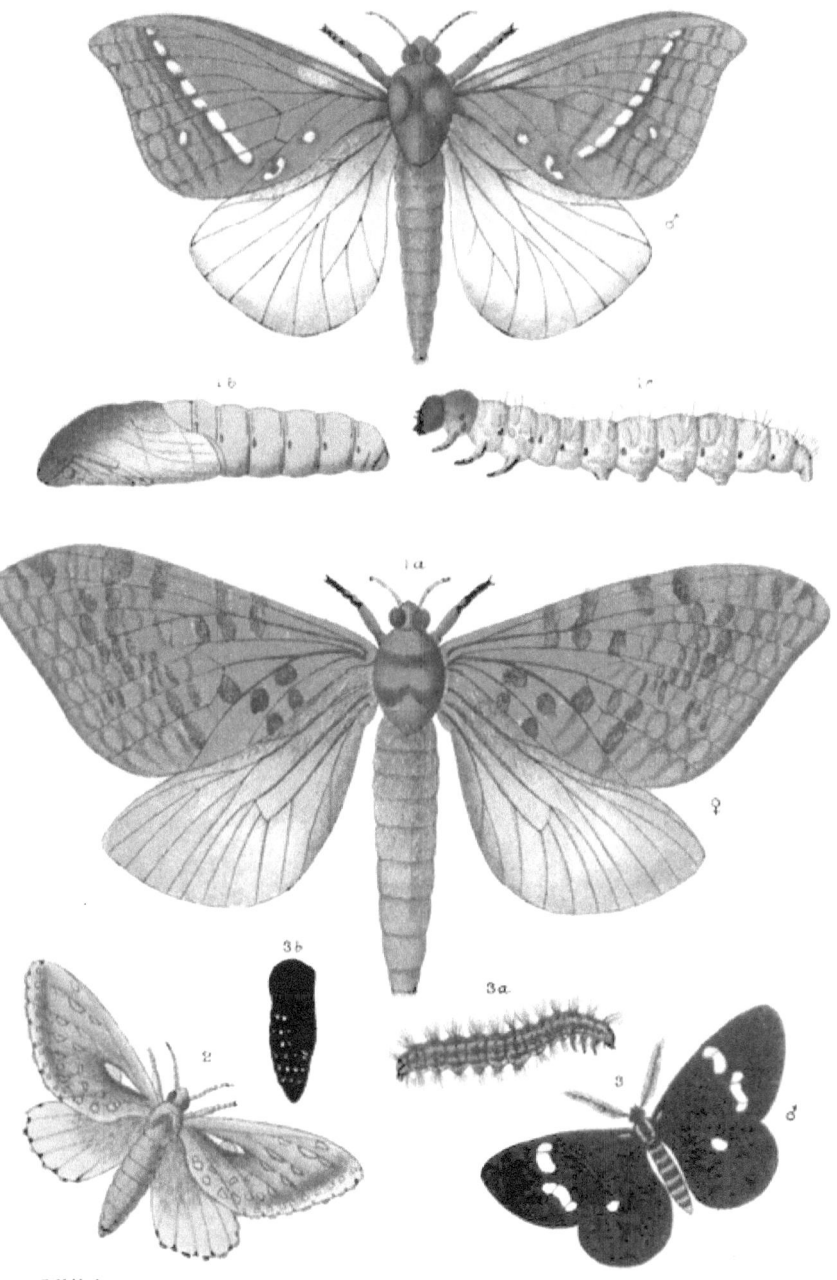

PLATE X.

Lepidoptera (*continued*).

Fig. 1.—Œceticus omnivorus ♂.
 „   1a.—   „   „   ♀.
 „   1b.—Larva.
 „   1c.—Male pupa.
 „   2.—Leucania atristriga ♂.
 „   3.—Mamestra composita ♂.
 „   3a.—Larva.
 „   4.—Heliothis armigera ♂.
 „   4a.—Larva.
 „   5.—Erana graminosa ♂.
 „   5a.—Larva.
 „   6.—Mamestra ustistriga, ♂.
 „   7.—   „   mutans ♂.
 „   7a.—Larva.
 „   7b.—Pupa.
 „   8.—Plusia eriosoma ♀.
 „   8a.—Larva.

Plate 7

PLATE XI.

Lepidoptera (*continued*).

Fig. 1.—Declana floccosa ♂.

  „ 1a.—Larva.

  „ 1b.—Declana floccosa, *var.* junctilinea ♂

  „ 2.—Chalastra pelurgata ♂.

  „ 2a.—   „   „   ♀.

  „ 2b.—Larva.

  „ 3.—Ploseria hemipteraria.

  „ 3a.—Larva.

  „ 4.—Ploseria alectoraria.

        (Larva at Plate XIII. fig. 7.)

  „ 5.—Sestra humeraria.

  „ 5a.—Larva.

  „ 6.—Sestra humeraria, *var.* (?)

  „ 7.—Selidosema panagrata ♂.

  „ 7a.—   „   „   ♀.

  „ 7b.—Larva.

  „ 8.—Selidosema dejectaria ♂.

  „ 8a.—   „   „   ♀.

  „ 8b.—Larva.

Plate XI

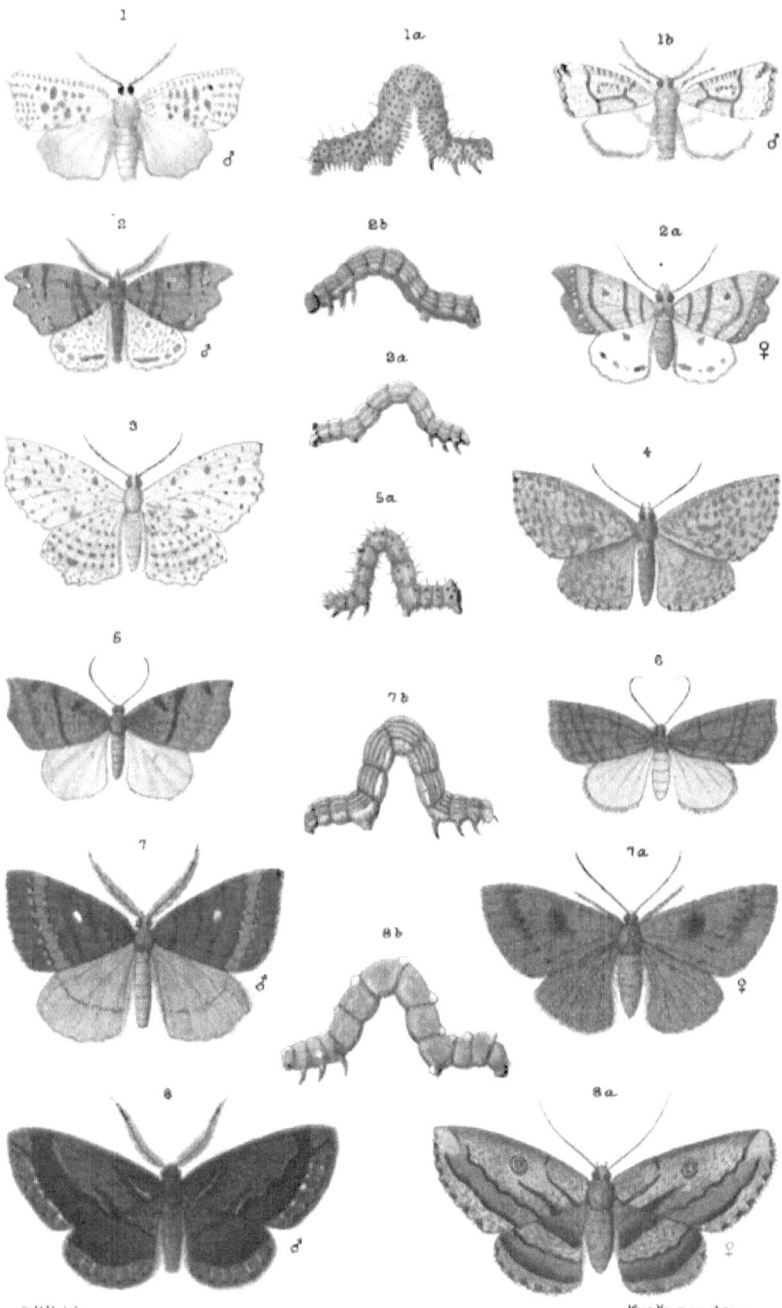

PLATE XII.

LEPIDOPTERA (*continued*).

Fig. 1.—Selidosema productata ♂.
 „ 1a.—   „   „   ♀.
 „ 1b.—Larva.
 „ 2.—Asthena schistaria.
 „ 2a.—Larva.
 „ 3.—Siculodes subfasciata.
 „ 3a.—Larva.
 „ 3b.—Section of stem showing enclosed
            pupa and aperture (*) through
            which moth escapes.
 „ 4.—Scoparia hemiplaca.
 „ 5.—Crambus flexuosellus.
 „ 6.—Ctenopseustis obliquana.
 „ 7.—Endrosis fenestrella.
 „ 7a.—Larva.
 „ 7b.—Pupa.
 „ 8.—Semiocosma platyptera.
 „ 8a.—Larva.
 „ 8b.—Pupa.

Plate XII

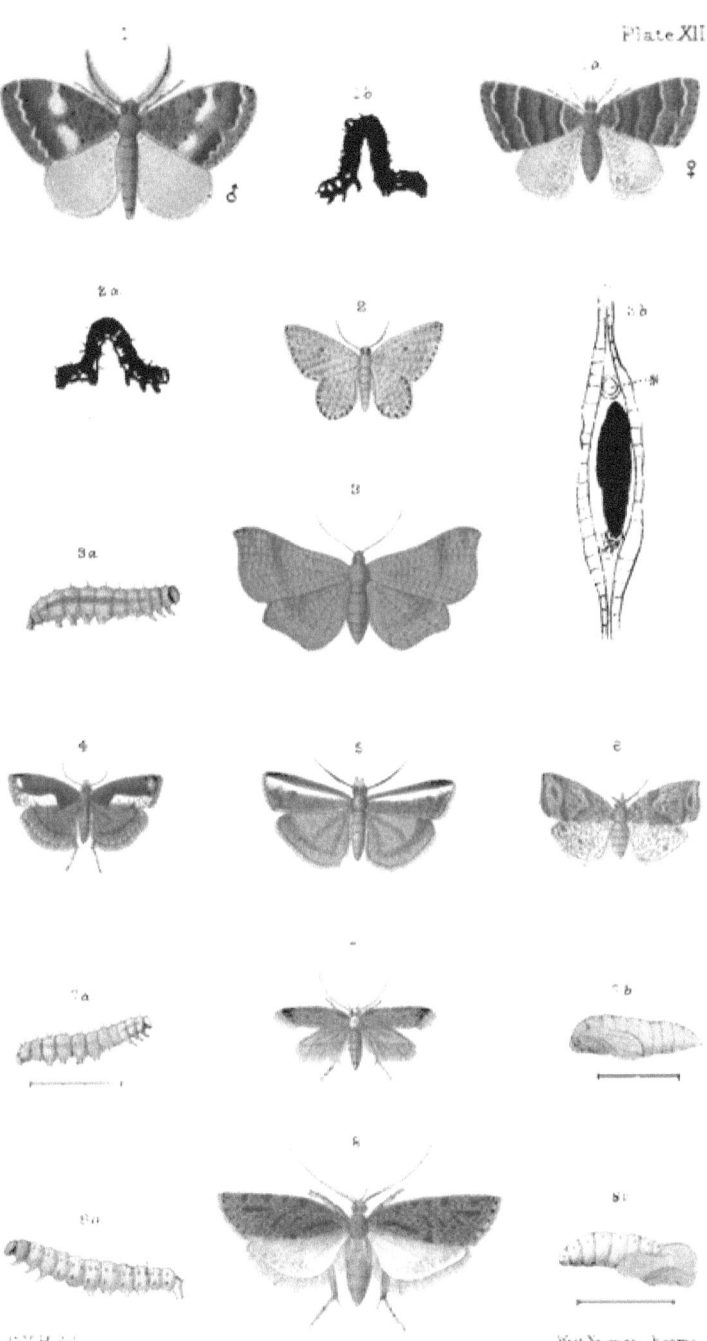

.

PLATE XIII.

Plate XIII.

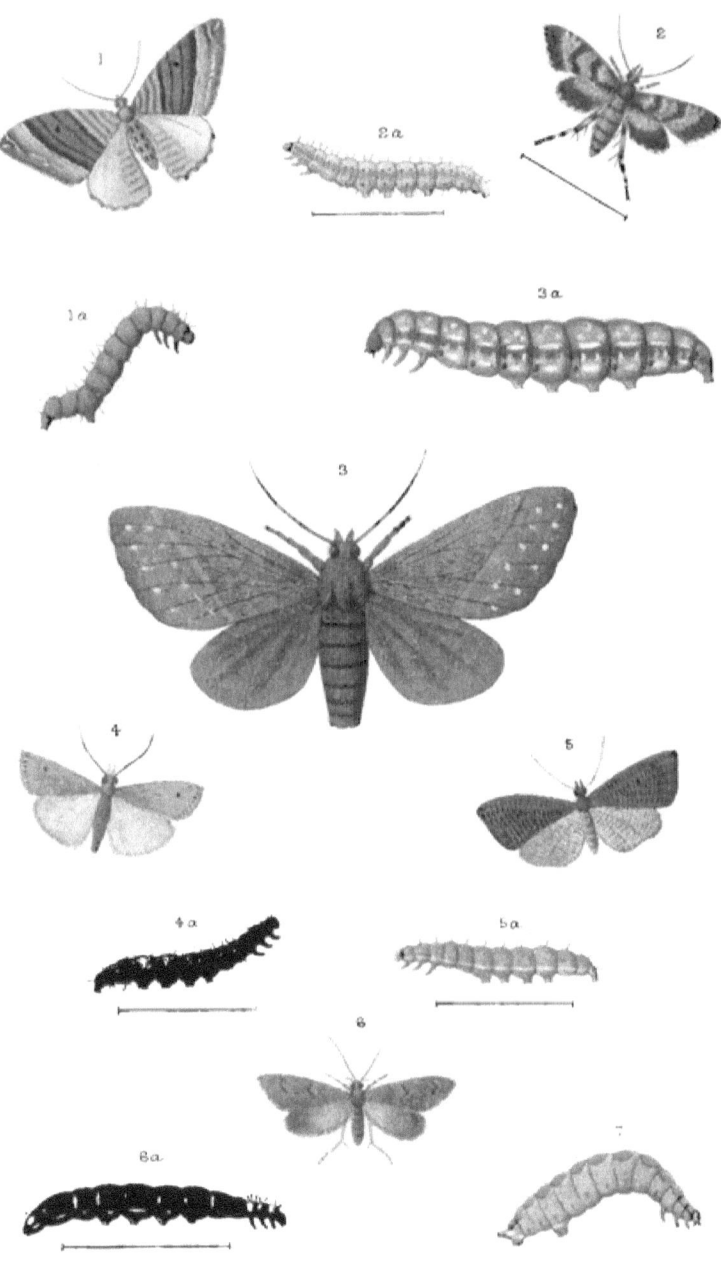

PLATE XIV.

NEUROPTERA.

Plate XI.

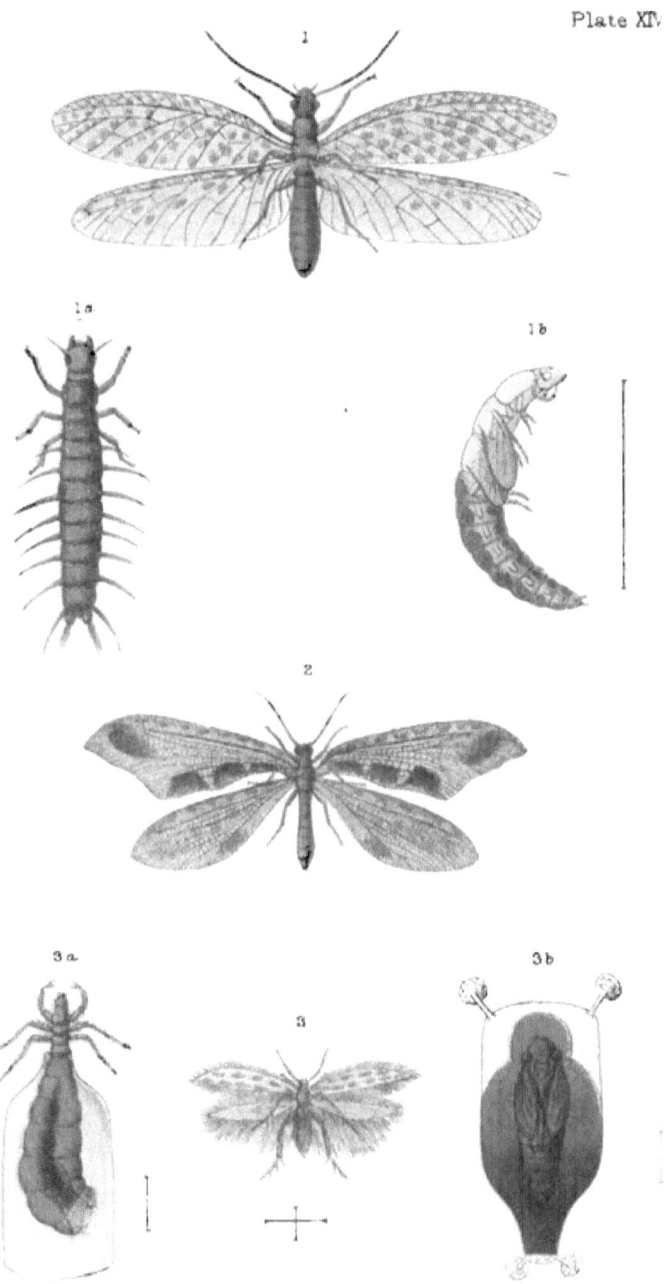

PLATE XV

ORTHOPTERA.

Plate XV

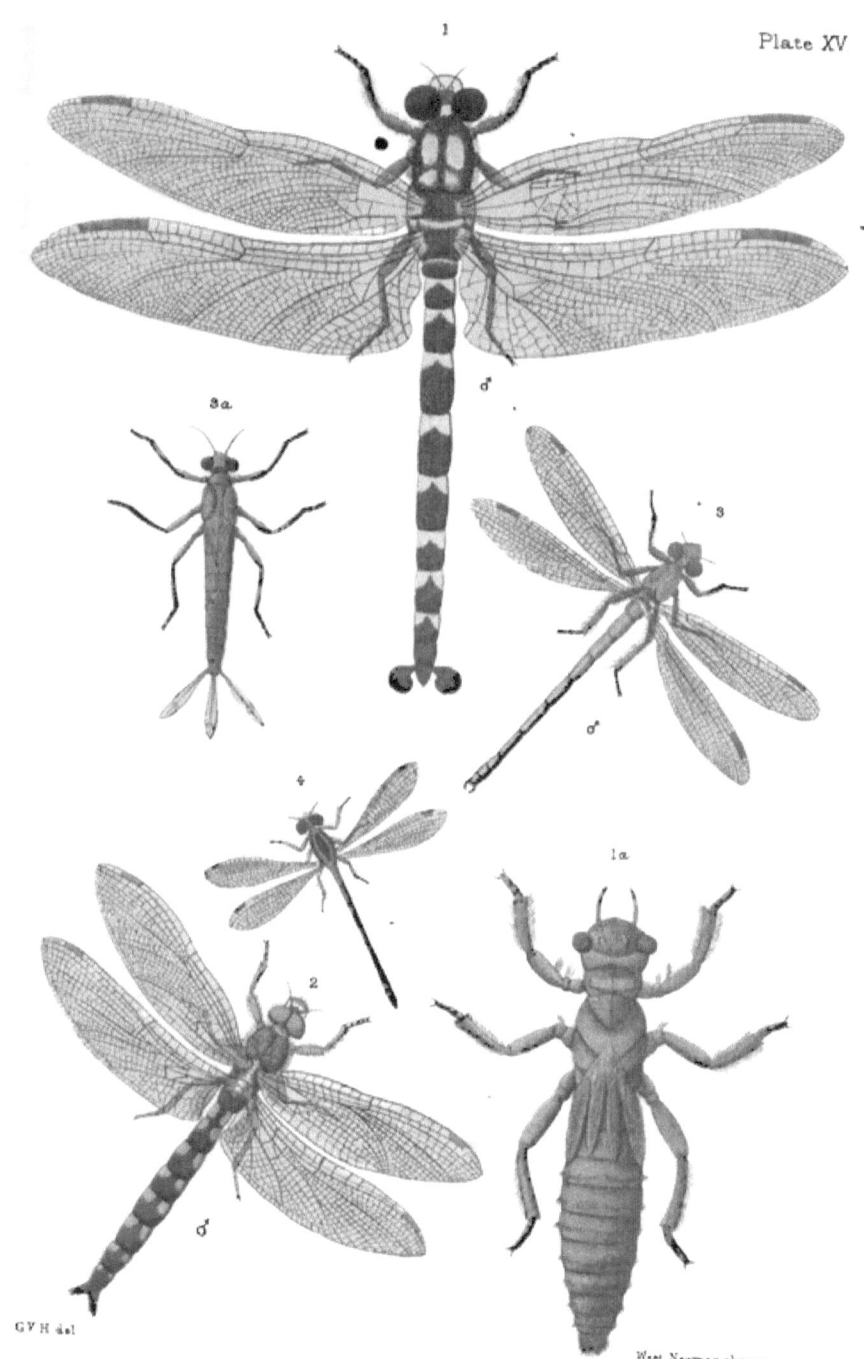

PLATE XVI.

ORTHOPTERA (*continued*).

Fig. 1.—Stolotermes ruficeps ♂.

 „ 1a.—Female.

 „ 1b.—Soldier.

 „ 1c.—Worker.

 „ 2.—Psocus zealandicus, n.s.

 „ 2a.—Larva.

 „ 3.—Stenoperla prasina.

 „ 3a.—Larva.

 „ 4.—Ephemera, n.s. (near Coloburus).

 „ 4a.—Larva.

Plate XVI

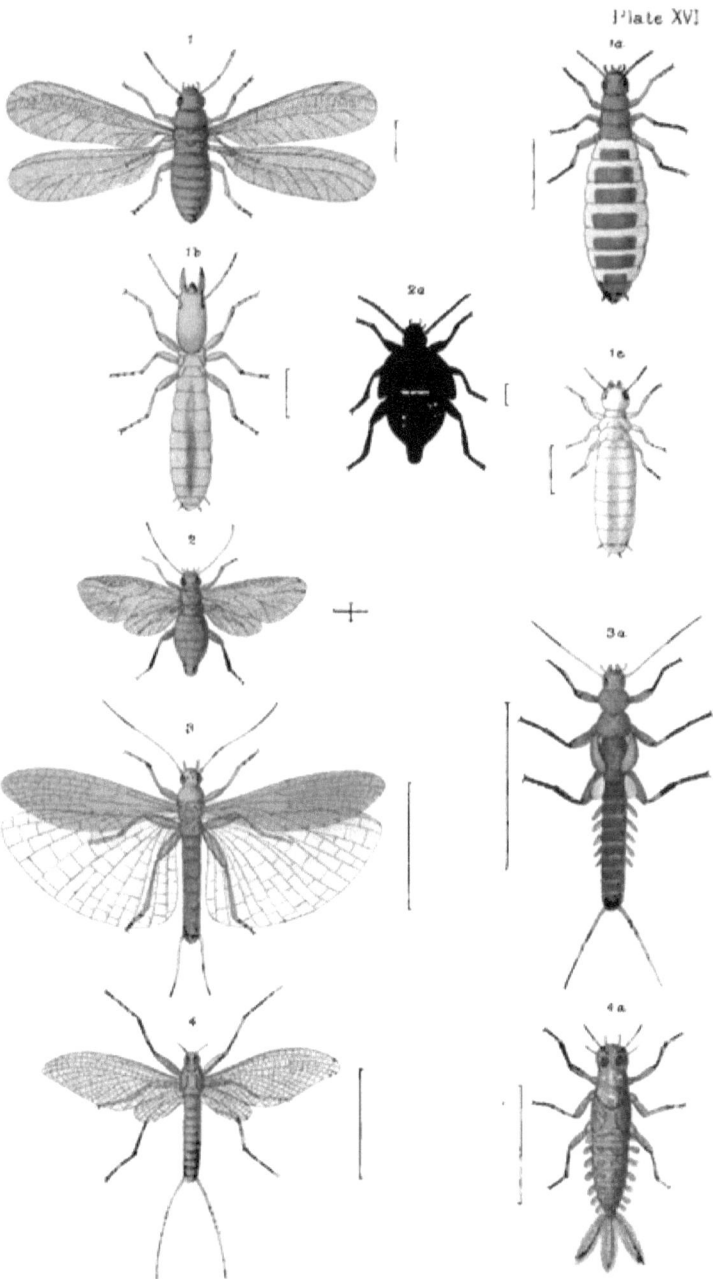

PLATE XVII.

ORTHOPTERA (*continued*).

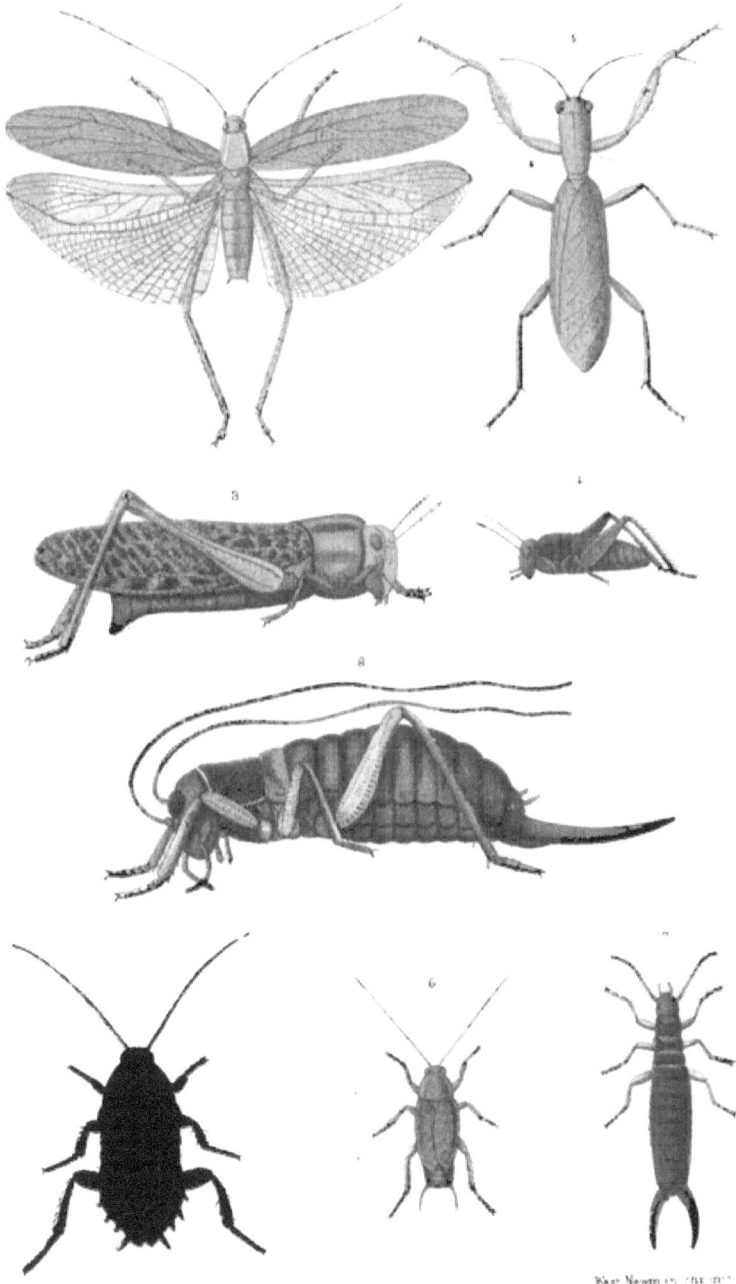

Plate XVI

West Newman chromo

PLATE XVIII.

Plate XVIII.

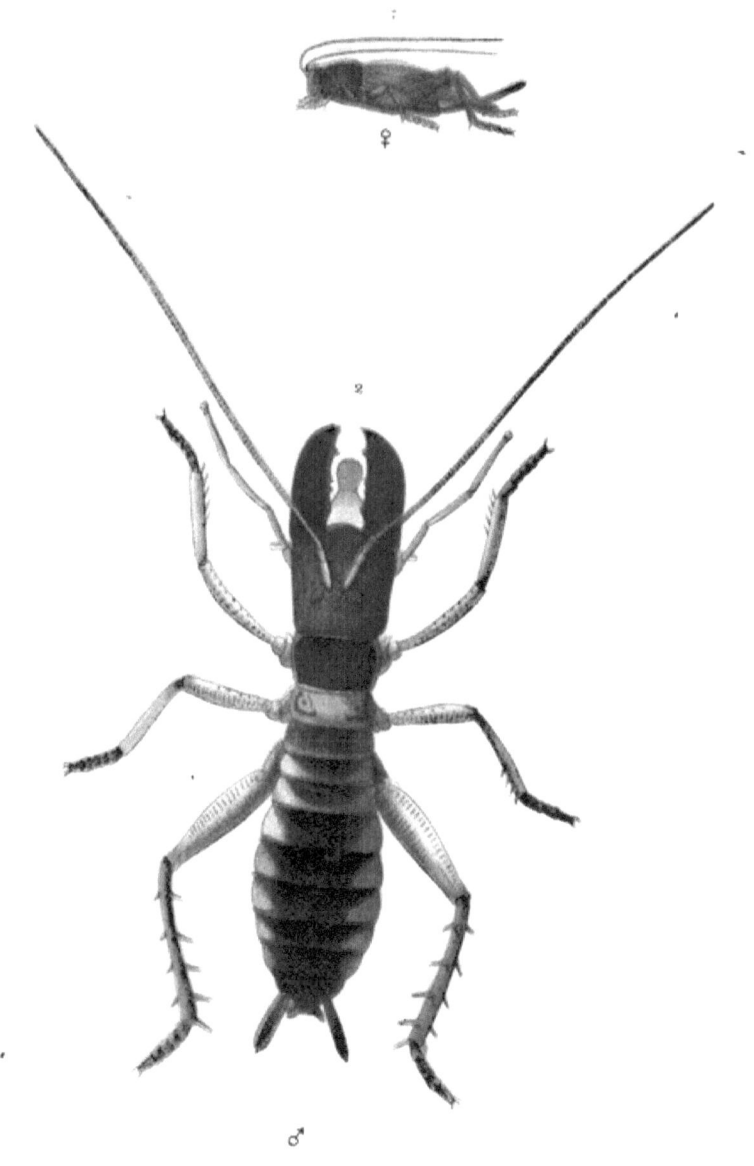

♀

2

♂

West, Newman chromo

PLATE XIX.

ORTHOPTERA (*concluded*).

Fig. 1.—Acanthoderus horridus.

Plate XIX.

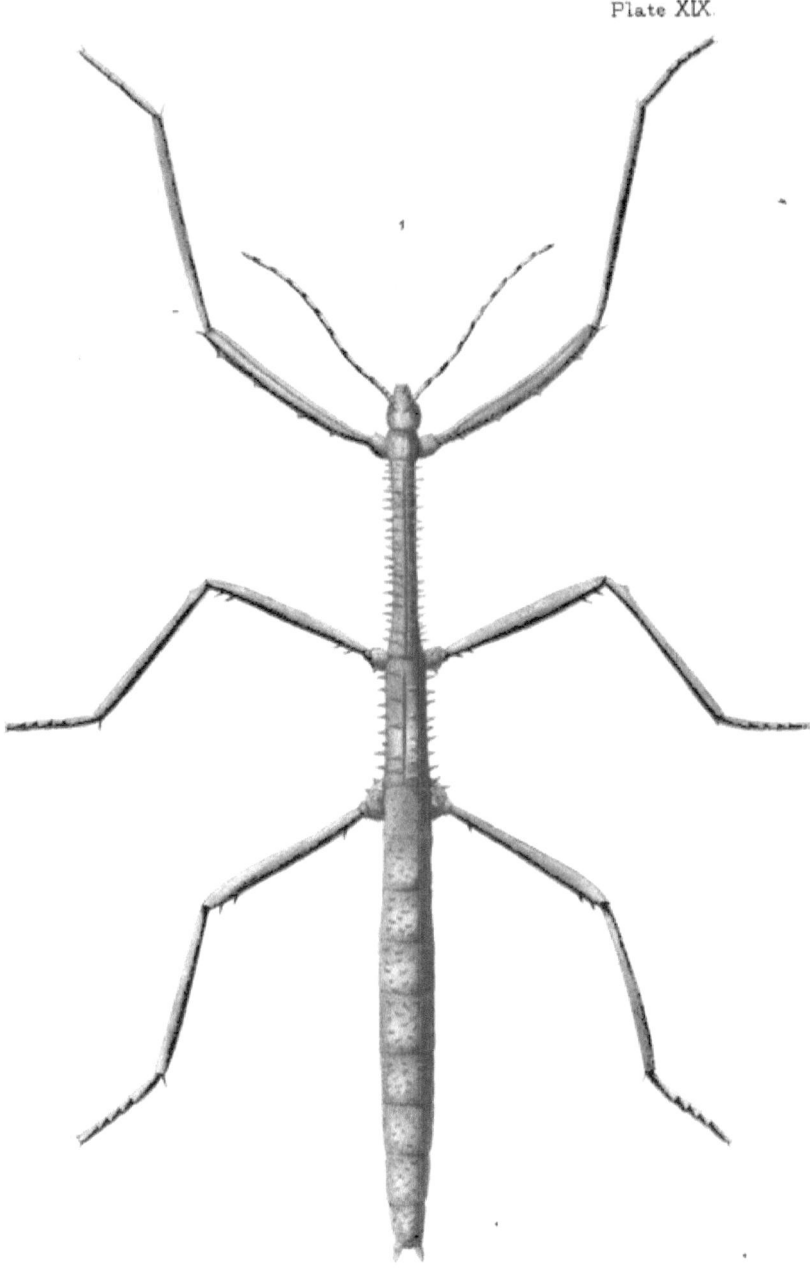

PLATE XX.

HEMIPTERA.

Fig. 1.—Cicada cingulata ♀.

  „ 1a.—Pupa.

  „ 2.—Cicada muta ♀.

  „ 3.— „ iolanthe, n.s.

  „ 3a.—Larva.

  „ 3b.—Pupa.

  „ 4.—Cœlostoma zealandicum ♂.

  „ 5.—Corixa zealandica.

  „ 6.—Cermatulus nasalis.

  „ 6a.—Larva.

Plate XX

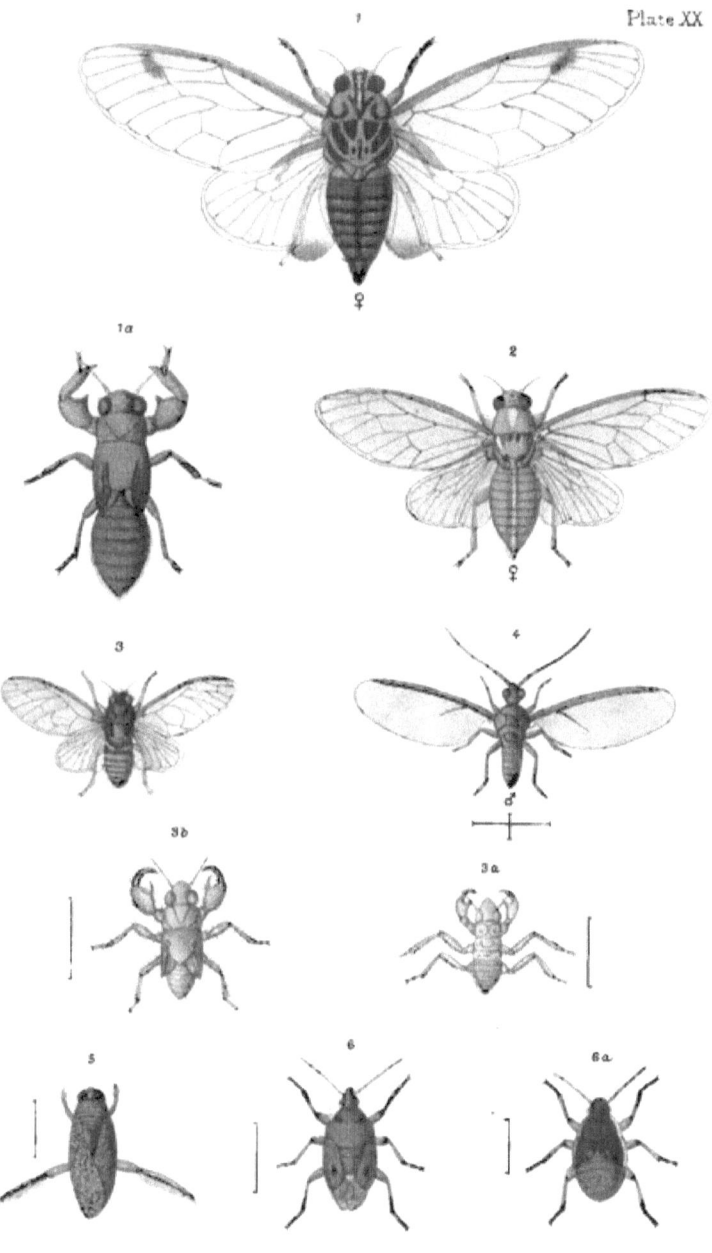